Springer Texts in Statistics

Series Editors:
Richard DeVeaux
Stephen E. Fienberg
Ingram Olkin

More information about this series at http://www.springer.com/series/417

Richard Durrett

Essentials of Stochastic Processes

Third Edition

Richard Durrett
Mathematics, Duke University
Durham, North Carolina, USA

ISSN 1431-875X   ISSN 2197-4136   (electronic)
Springer Texts in Statistics
ISBN 978-3-319-83331-6   ISBN 978-3-319-45614-0   (eBook)
DOI 10.1007/978-3-319-45614-0

Printed on acid-free paper

This Springer imprint is published by Springer Nature
The registered company is Springer International Publishing AG
The registered company address is: Gewerbestrasse 11, 6330 Cham, Switzerland

Preface

Between the first undergraduate course in probability and the first graduate course that uses measure theory, there are a number of courses that teach *stochastic processes* to students with many different interests and with varying degrees of mathematical sophistication. To allow readers (and instructors) to choose their own level of detail, many of the proofs begin with a nonrigorous answer to the question "Why is this true?" followed by a *proof* that fills in the missing details. As it is possible to drive a car without knowing about the working of the internal combustion engine, it is also possible to apply the theory of Markov chains without knowing the details of the proofs. It is my personal philosophy that probability theory was developed to solve problems, so most of our effort will be spent on analyzing examples. Readers who want to master the subject will have to do more than a few of the 20 dozen carefully chosen exercises.

This book began as notes I typed in the spring of 1997 as I was teaching ORIE 361 at Cornell for the second time. In spring of 2009, the mathematics department at Cornell introduced its own version of this course, MATH 474. This started me on the task of preparing the second edition. The plan was to have this finished in spring of 2010 after the second time I taught the course, but when May rolled around, completing the book lost out to getting ready to move to Durham after 25 years in Ithaca. In the fall of 2011, I taught Duke's version of the course, MATH 216, to 20 undergrads and 12 graduate students, and over the Christmas break, the second edition was completed.

The second edition differs substantially from the first, though curiously the length and the number of problems have remained roughly constant. Throughout the book, there are many new examples and problems, with solutions that use the TI-83 to eliminate the tedious details of solving linear equations by hand. The Markov chains chapter has been reorganized. The chapter on Poisson processes has moved up from third to second and is now followed by a treatment of the closely related topic of renewal theory. Continuous time Markov chains remain fourth, with a new section on exit distributions and hitting times and reduced coverage of queueing networks. Martingales, a difficult subject for students at this level, now comes fifth, in order to set the stage for their use in a new sixth chapter on mathematical finance. The treatment of finance expands the two sections of the previous treatment to include American options and the capital asset pricing model. Brownian motion makes a cameo appearance in the discussion of the Black-Scholes theorem, but in contrast to the previous edition, it is not discussed in detail.

The changes in the third edition are more modest. When I taught the course in the fall of 2015, the 47 students were rather vocal about the shortcomings of the book, so there are a number of changes throughout. The only major structural change is that the proofs in chapter have been reorganized to separate the useful idea of returns to a fixed state from the more mysterious coupling proof of the convergence theorem. If you find new ones, email: rtd@math.duke.edu.

Durham, NC, USA   Rick Durrett

Contents

Chapter 1
Markov Chains

1.1 Definitions and Examples

The importance of Markov chains comes from two facts: (i) there are a large number of physical, biological, economic, and social phenomena that can be modeled in this way, and (ii) there is a well-developed theory that allows us to do computations. We begin with a famous example, then describe the property that is the defining feature of Markov chains

Example 1.1 (Gambler's Ruin). Consider a gambling game in which on any turn you win \$1 with probability $p = 0.4$ or lose \$1 with probability $1 - p = 0.6$. Suppose further that you adopt the rule that you quit playing if your fortune reaches \$N. Of course, if your fortune reaches \$0 the casino makes you stop.

Let X_n be the amount of money you have after n plays. Your fortune, X_n has the "Markov property." In words, this means that given the current state, X_n, any other information about the past is irrelevant for predicting the next state X_{n+1}. To check this for the gambler's ruin chain, we note that if you are still playing at time n, i.e., your fortune $X_n = i$ with $0 < i < N$, then for any possible history of your wealth $i_{n-1}, i_{n-2}, \ldots i_1, i_0$

$$P(X_{n+1} = i + 1 | X_n = i, X_{n-1} = i_{n-1}, \ldots X_0 = i_0) = 0.4$$

since to increase your wealth by one unit you have to win your next bet. Here we have used $P(B|A)$ for the conditional probability of the event B given that A occurs. Recall that this is defined by

$$P(B|A) = \frac{P(B \cap A)}{P(A)}$$

If you need help with this notion, see Sect. A.1 of the appendix.

© Springer International Publishing Switzerland 2016
R. Durrett, *Essentials of Stochastic Processes*, Springer Texts in Statistics,
DOI 10.1007/978-3-319-45614-0_1

Turning now to the formal definition, we say that X_n is a discrete time **Markov chain** with **transition matrix** $p(i,j)$ if for any $j, i, i_{n-1}, \ldots i_0$

$$P(X_{n+1} = j | X_n = i, X_{n-1} = i_{n-1}, \ldots, X_0 = i_0) = p(i,j) \qquad (1.1)$$

Here and in what follows, **boldface** indicates a word or phrase that is being defined or explained.

_ Equation (1.1) explains what we mean when we say that "given the current state X_n, any other information about the past is irrelevant for predicting X_{n+1}." In formulating (1.1) we have restricted our attention to the **temporally homogeneous** case in which the **transition probability**

$$p(i,j) = P(X_{n+1} = j | X_n = i)$$

does not depend on the time n.

Intuitively, the transition probability gives the rules of the game. It is the basic information needed to describe a Markov chain. In the case of the gambler's ruin chain, the transition probability has

$$p(i, i+1) = 0.4, \quad p(i, i-1) = 0.6, \quad \text{if } 0 < i < N$$
$$p(0,0) = 1 \qquad p(N,N) = 1$$

When $N = 5$ the matrix is

	0	1	2	3	4	5
0	1.0	0	0	0	0	0
1	0.6	0	0.4	0	0	0
2	0	0.6	0	0.4	0	0
3	0	0	0.6	0	0.4	0
4	0	0	0	0.6	0	0.4
5	0	0	0	0	0	1.0

or the chain be represented pictorially as

Example 1.2 (Ehrenfest Chain). This chain originated in physics as a model for two cubical volumes of air connected by a small hole. In the mathematical version, we have two "urns," i.e., two of the exalted trash cans of probability theory, in which there are a total of N balls. We pick one of the N balls at random and move it to the other urn.

Let X_n be the number of balls in the "left" urn after the nth draw. It should be clear that X_n has the Markov property; i.e., if we want to guess the state at time $n+1$, then the current number of balls in the left urn X_n is the only relevant information from the observed sequence of states $X_n, X_{n-1}, \ldots X_1, X_0$. To check this we note that

$$P(X_{n+1} = i + 1 | X_n = i, X_{n-1} = i_{n-1}, \ldots X_0 = i_0) = (N - i)/N$$

since to increase the number we have to pick one of the $N - i$ balls in the other urn. The number can also decrease by 1 with probability i/N. In symbols, we have computed that the transition probability is given by

$$p(i, i + 1) = (N - i)/N, \quad p(i, i - 1) = i/N \qquad \text{for } 0 \le i \le N$$

with $p(i, j) = 0$ otherwise. When $N = 4$, for example, the matrix is

	0	1	2	3	4
0	0	1	0	0	0
1	1/4	0	3/4	0	0
2	0	2/4	0	2/4	0
3	0	0	3/4	0	1/4
4	0	0	0	1	0

In the first two examples we began with a verbal description and then wrote down the transition probabilities. However, one more commonly describes a Markov chain by writing down a transition probability $p(i, j)$ with

(i) $p(i, j) \ge 0$, since they are probabilities.
(ii) $\sum_j p(i, j) = 1$, since when $X_n = i$, X_{n+1} will be in some state j.

The equation in (ii) is read "sum $p(i, j)$ over all possible values of j." In words the last two conditions say: the entries of the matrix are nonnegative and each ROW of the matrix sums to 1.

Any matrix with properties (i) and (ii) gives rise to a Markov chain, X_n. To construct the chain we can think of playing a board game. When we are in state i, we roll a die (or generate a random number on a computer) to pick the next state, going to j with probability $p(i, j)$.

Example 1.3 (Weather Chain). Let X_n be the weather on day n in Ithaca, NY, which we assume is either: 1 = *rainy*, or 2 = *sunny*. Even though the weather is not exactly a Markov chain, we can propose a Markov chain model for the weather by writing down a transition probability

$$
\begin{array}{cc}
 & 1 \; 2 \\
1 & .6 \; .4 \\
2 & .2 \; .8
\end{array}
$$

The table says, for example, the probability a rainy day (state 1) is followed by a sunny day (state 2) is $p(1,2) = 0.4$. A typical question of interest is

Q. What is the long-run fraction of days that are sunny?

Example 1.4 (Social Mobility). Let X_n be a family's social class in the nth generation, which we assume is either $1 = lower$, $2 = middle$, or $3 = upper$. In our simple version of sociology, changes of status are a Markov chain with the following transition probability

$$
\begin{array}{cccc}
 & 1 \; 2 \; 3 \\
1 & .7 \; .2 \; .1 \\
2 & .3 \; .5 \; .2 \\
3 & .2 \; .4 \; .4
\end{array}
$$

Q. Do the fractions of people in the three classes approach a limit?

Example 1.5 (Brand Preference). Suppose there are three types of laundry detergent, 1, 2, and 3, and let X_n be the brand chosen on the nth purchase. Customers who try these brands are satisfied and choose the same thing again with probabilities 0.8, 0.6, and 0.4, respectively. When they change they pick one of the other two brands at random. The transition probability is

$$
\begin{array}{cccc}
 & 1 \; 2 \; 3 \\
1 & .8 \; .1 \; .1 \\
2 & .2 \; .6 \; .2 \\
3 & .3 \; .3 \; .4
\end{array}
$$

Q. Do the market shares of the three product stabilize?

Example 1.6 (Inventory Chain). We will consider the consequences of using an s, S inventory control policy. That is, when the stock on hand at the end of the day falls to s or below, we order enough to bring it back up to S. For simplicity, we suppose happens at the beginning of the next day. Let X_n be the amount of stock on hand at the end of day n and D_{n+1} be the demand on day $n+1$. Introducing notation for the **positive part** of a real number,

$$
x^+ = \max\{x, 0\} = \begin{cases} x & \text{if } x > 0 \\ 0 & \text{if } x \le 0 \end{cases}
$$

then we can write the chain in general as

$$X_{n+1} = \begin{cases} (X_n - D_{n+1})^+ & \text{if } X_n > s \\ (S - D_{n+1})^+ & \text{if } X_n \le s \end{cases}$$

In words, if $X_n > s$ we order nothing and begin the day with X_n units. If the demand $D_{n+1} \le X_n$, we end the day with $X_{n+1} = X_n - D_{n+1}$. If the demand $D_{n+1} > X_n$, we end the day with $X_{n+1} = 0$. If $X_n \le s$, then we begin the day with S units, and the reasoning is the same as in the previous case.

Suppose now that an electronics store sells a video game system and uses an inventory policy with $s = 1, S = 5$. That is, if at the end of the day, the number of units they have on hand is 1 or 0, they order enough new units so their total on hand at the beginning of the next day is 5. If we assume that

for $k =$	0	1	2	3
$P(D_{n+1} = k)$	.3	.4	.2	.1

then we have the following transition matrix:

	0	1	2	3	4	5
0	0	0	.1	.2	.4	.3
1	0	0	.1	.2	.4	.3
2	.3	.4	.3	0	0	0
3	.1	.2	.4	.3	0	0
4	0	.1	.2	.4	.3	0
5	0	0	.1	.2	.4	.3

To explain the entries, we note that when $X_n \ge 3$ then $X_n - D_{n+1} \ge 0$. When $X_{n+1} = 2$ this is almost true but $p(2, 0) = P(D_{n+1} = 2 \text{ or } 3)$. When $X_n = 1$ or 0 we start the day with five units so the end result is the same as when $X_n = 5$.

In this context we might be interested in:

Q. Suppose we make \$12 profit on each unit sold but it costs \$2 a day to store items. What is the long-run profit per day of this inventory policy? How do we choose s and S to maximize profit?

Example 1.7 (Repair Chain). A machine has three critical parts that are subject to failure, but can function as long as two of these parts are working. When two are broken, they are replaced and the machine is back to working order the next day. To formulate a Markov chain model we declare its state space to be the parts that are broken $\{0, 1, 2, 3, 12, 13, 23\}$. If we assume that parts 1, 2, and 3 fail with probabilities .01, .02, and .04, but no two parts fail on the same day, then we arrive at the following transition matrix:

	0	1	2	3	12	13	23
0	.93	.01	.02	.04	0	0	0
1	0	.94	0	0	.02	.04	0
2	0	0	.95	0	.01	0	.04
3	0	0	0	.97	0	.01	.02
12	1	0	0	0	0	0	0
13	1	0	0	0	0	0	0
23	1	0	0	0	0	0	0

If we own a machine like this, then it is natural to ask about the long-run cost per day to operate it. For example, we might ask:

Q. If we are going to operate the machine for 1800 days (about 5 years), then how many parts of types 1, 2, and 3 will we use?

Example 1.8 (Branching Processes). These processes arose from Francis Galton's statistical investigation of the extinction of family names. Consider a population in which each individual in the nth generation independently gives birth, producing k children (who are members of generation $n + 1$) with probability p_k. In Galton's application only male children count since only they carry on the family name.

To define the Markov chain, note that the number of individuals in generation n, X_n, can be any nonnegative integer, so the state space is $\{0, 1, 2, \ldots\}$. If we let $Y_1, Y_2, \ldots$ be independent random variables with $P(Y_m = k) = p_k$, then we can write the transition probability as

$$p(i,j) = P(Y_1 + \cdots + Y_i = j) \quad \text{for } i > 0 \text{ and } j \geq 0$$

When there are no living members of the population, no new ones can be born, so $p(0,0) = 1$.

Galton's question, originally posed in the Educational Times of 1873, is

Q. What is the probability that the line of a man becomes extinct?, i.e., the branching process becomes absorbed at 0?

Reverend Henry William Watson replied with a solution. Together, they then wrote an 1874 paper entitled *On the probability of extinction of families*. For this reason, these chains are often called Galton–Watson processes.

Example 1.9 (Wright–Fisher Model). Thinking of a population of $N/2$ diploid individuals who have two copies of each of their chromosomes, or of N haploid individuals who have one copy, we consider a fixed population of N genes that can be one of two types: A or a. In the simplest version of this model the population at time $n + 1$ is obtained by drawing with replacement from the population at time n.

In this case, if we let X_n be the number of A alleles at time n, then X_n is a Markov chain with transition probability

$$p(i,j) = \binom{N}{j}\left(\frac{i}{N}\right)^j\left(1 - \frac{i}{N}\right)^{N-j}$$

since the right-hand side is the binomial distribution for N independent trials with success probability i/N.

In this model the states $x = 0$ and N that correspond to fixation of the population in the all a or all A states are **absorbing states**, that is, $p(x, x) = 1$. So it is natural to ask:

Q1. Starting from i of the A alleles and $N - i$ of the a alleles, what is the probability that the population fixates in the all A state?

To make this simple model more realistic we can introduce the possibility of mutations: an A that is drawn ends up being an a in the next generation with probability u, while an a that is drawn ends up being an A in the next generation with probability v. In this case the probability an A is produced by a given draw is

$$\rho_i = \frac{i}{N}(1 - u) + \frac{N - i}{N}v$$

but the transition probability still has the binomial form

$$p(i,j) = \binom{N}{j}(\rho_i)^j(1 - \rho_i)^{N-j}$$

If u and v are both positive, then 0 and N are no longer absorbing states, so we ask:

Q2. Does the genetic composition settle down to an equilibrium distribution as time $t \to \infty$?

As the next example shows it is easy to extend the notion of a Markov chain to cover situations in which the future evolution is independent of the past when we know the last two states.

Example 1.10 (Two-Stage Markov Chains). In a Markov chain the distribution of X_{n+1} only depends on X_n. This can easily be generalized to case in which the distribution of X_{n+1} only depends on (X_n, X_{n-1}). For a concrete example consider a basketball player who makes a shot with the following probabilities:

1/2 if he has missed the last two times
2/3 if he has hit one of his last two shots
3/4 if he has hit both of his last two shots

To formulate a Markov chain to model his shooting, we let the states of the process be the outcomes of his last two shots: $\{HH, HM, MH, MM\}$ where M is short for miss and H for hit. The transition probability is

$$
\begin{array}{c|cccc}
 & \textbf{HH} & \textbf{HM} & \textbf{MH} & \textbf{MM} \\
\hline
\textbf{HH} & 3/4 & 1/4 & 0 & 0 \\
\textbf{HM} & 0 & 0 & 2/3 & 1/3 \\
\textbf{MH} & 2/3 & 1/3 & 0 & 0 \\
\textbf{MM} & 0 & 0 & 1/2 & 1/2 \\
\end{array}
$$

To explain suppose the state is HM, i.e., $X_{n-1} = H$ and $X_n = M$. In this case the next outcome will be H with probability 2/3. When this occurs the next state will be $(X_n, X_{n+1}) = (M, H)$ with probability 2/3. If he misses an event of probability 1/3, $(X_n, X_{n+1}) = (M, M)$.

The Hot Hand is a phenomenon known to most people who play or watch basketball. After making a couple of shots, players are thought to "get into a groove" so that subsequent successes are more likely. Purvis Short of the Golden State Warriors describes this more poetically as

> You're in a world all your own. It's hard to describe. But the basket seems to be so wide. No matter what you do, you know the ball is going to go in.

Unfortunately for basketball players, data collected by Gliovich et al. (1985) shows that this is a misconception. The next table gives data for the conditional probability of hitting a shot after missing the last three, missing the last two, ... hitting the last three, for nine players of the Philadelphia 76ers: Darryl Dawkins (403), Maurice Cheeks (339), Steve Mix (351), Bobby Jones (433), Clint Richardson (248), Julius Erving (884), Andrew Toney (451), Caldwell Jones (272), and Lionel Hollins (419). The numbers in parentheses are the number of shots for each player.

$P(H|3M)$	$P(H|2M)$	$P(H|1M)$	$P(H|1H)$	$P(H|2H)$	$P(H|3H)$
.88	.73	.71	.57	.58	.51
.77	.60	.60	.55	.54	.59
.70	.56	.52	.51	.48	.36
.61	.58	.58	.53	.47	.53
.52	.51	.51	.53	.52	.48
.50	.47	.56	.49	.50	.48
.50	.48	.47	.45	.43	.27
.52	.53	.51	.43	.40	.34
.50	.49	.46	.46	.46	.32

In fact, the data supports the opposite assertion: after missing a player will hit more frequently.

1.2 Multistep Transition Probabilities

The transition probability $p(i,j) = P(X_{n+1} = j|X_n = i)$ gives the probability of going from i to j in one step. Our goal in this section is to compute the probability of going from i to j in $m > 1$ steps:

$$p^m(i,j) = P(X_{n+m} = j|X_n = i)$$

As the notation may already suggest, p^m will turn out to be the mth power of the transition matrix, see Theorem 1.1.

To warm up, we recall the transition probability of the social mobility chain:

$$
\begin{array}{c|ccc}
 & 1 & 2 & 3 \\
\hline
1 & .7 & .2 & .1 \\
2 & .3 & .5 & .2 \\
3 & .2 & .4 & .4 \\
\end{array}
$$

and consider the following concrete question:

Q1. *Your parents were middle class (state 2). What is the probability that you are in the upper class (state 3) but your children are lower class (state 1)?*

Solution. Intuitively, the Markov property implies that starting from state 2 the probability of jumping to 3 and then to 1 is given by

$$p(2,3)p(3,1)$$

To get this conclusion from the definitions, we note that using the definition of conditional probability,

$$P(X_2 = 1, X_1 = 3|X_0 = 2) = \frac{P(X_2 = 1, X_1 = 3, X_0 = 2)}{P(X_0 = 2)}$$

$$= \frac{P(X_2 = 1, X_1 = 3, X_0 = 2)}{P(X_1 = 3, X_0 = 2)} \cdot \frac{P(X_1 = 3, X_0 = 2)}{P(X_0 = 2)}$$

$$= P(X_2 = 1|X_1 = 3, X_0 = 2) \cdot P(X_1 = 3|X_0 = 2)$$

By the Markov property (1.1) the last expression is

$$P(X_2 = 1|X_1 = 3) \cdot P(X_1 = 3|X_0 = 2) = p(2,3)p(3,1)$$

Moving on to the real question:

Q2. *What is the probability your children are lower class (1) given your parents were middle class (2)?*

Solution. To do this we simply have to consider the three possible states for your class and use the solution of the previous problem.

$$P(X_2 = 1|X_0 = 2) = \sum_{k=1}^{3} P(X_2 = 1, X_1 = k|X_0 = 2) = \sum_{k=1}^{3} p(2,k)p(k,1)$$

$$= (.3)(.7) + (.5)(.3) + (.2)(.2) = .21 + .15 + .04 = .21$$

There is nothing special here about the states 2 and 1 here. By the same reasoning,

$$P(X_2 = j|X_0 = i) = \sum_{k=1}^{3} p(i,k)\,p(k,j)$$

The right-hand side of the last equation gives the (i,j)th entry of the matrix p is multiplied by itself.

To explain this, we note that to compute $p^2(2,1)$ we multiplied the entries of the second row by those in the first column:

$$\begin{pmatrix} . & . & . \\ .3 & .5 & .2 \\ . & . & . \end{pmatrix} \begin{pmatrix} .7 & . & . \\ .3 & . & . \\ .2 & . & . \end{pmatrix} = \begin{pmatrix} . & . & . \\ .40 & . & . \\ . & . & . \end{pmatrix}$$

If we wanted $p^2(1,3)$ we would multiply the first row by the third column:

$$\begin{pmatrix} .7 & .2 & .1 \\ . & . & . \\ . & . & . \end{pmatrix} \begin{pmatrix} . & . & .1 \\ . & . & .2 \\ . & . & .4 \end{pmatrix} = \begin{pmatrix} . & . & .15 \\ . & . & . \\ . & . & . \end{pmatrix}$$

When all of the computations are done we have

$$\begin{pmatrix} .7 & .2 & .1 \\ .3 & .5 & .2 \\ .2 & .4 & .4 \end{pmatrix} \begin{pmatrix} .7 & .2 & .1 \\ .3 & .5 & .2 \\ .2 & .4 & .4 \end{pmatrix} = \begin{pmatrix} .57 & .28 & .15 \\ .40 & .39 & .21 \\ .34 & .40 & .26 \end{pmatrix}$$

All of this becomes much easier if we use a scientific calculator like the T1-83. Using 2nd-MATRIX we can access a screen with NAMES, MATH, EDIT at the top. Selecting EDIT we can enter the matrix into the computer as say [A]. Then selecting the NAMES we can enter $[A] \wedge 2$ on the computation line to get A^2. If we use this procedure to compute A^{20}, we get a matrix with three rows that agree in the first six decimal places with

.468085   .340425   .191489

Later we will see that as $n \to \infty$, p^n converges to a matrix with all three rows equal to $(22/47, 16/47, 9/47)$.

To explain our interest in p^m we will now prove:

Theorem 1.1. *The m step transition probability $P(X_{n+m} = j|X_n = i)$ is the mth power of the transition matrix p.*

The key ingredient in proving this is the **Chapman–Kolmogorov equation**

$$p^{m+n}(i,j) = \sum_k p^m(i,k)\, p^n(k,j) \tag{1.2}$$

Once this is proved, Theorem 1.1 follows, since taking $n = 1$ in (1.2), we see that

$$p^{m+1}(i,j) = \sum_k p^m(i,k)\, p(k,j)$$

That is, the $m + 1$ step transition probability is the m step transition probability times p. Theorem 1.1 now follows.

Why is (1.2) true? To go from i to j in $m + n$ steps, we have to go from i to some state k in m steps and then from k to j in n steps. The Markov property implies that the two parts of our journey are independent. $\qquad\square$

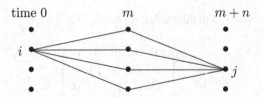

Proof of (1.2). We do this by combining the solutions of Q1 and Q2. Breaking things down according to the state at time m,

$$P(X_{m+n} = j|X_0 = i) = \sum_k P(X_{m+n} = j, X_m = k|X_0 = i)$$

Using the definition of conditional probability as in the solution of Q1,

$$P(X_{m+n} = j, X_m = k|X_0 = i) = \frac{P(X_{m+n} = j, X_m = k, X_0 = i)}{P(X_0 = i)}$$

$$= \frac{P(X_{m+n} = j, X_m = k, X_0 = i)}{P(X_m = k, X_0 = i)} \cdot \frac{P(X_m = k, X_0 = i)}{P(X_0 = i)}$$

$$= P(X_{m+n} = j|X_m = k, X_0 = i) \cdot P(X_m = k|X_0 = i)$$

By the Markov property (1.1) the last expression is

$$= P(X_{m+n} = j|X_m = k) \cdot P(X_m = k|X_0 = i) = p^m(i,k)p^n(k,j)$$

and we have proved (1.2).   □

Having established (1.2), we now return to computations.

Example 1.11 (Gambler's Ruin). Suppose for simplicity that $N = 4$ in Example 1.1, so that the transition probability is

	0	1	2	3	4
0	1.0	0	0	0	0
1	0.6	0	0.4	0	0
2	0	0.6	0	0.4	0
3	0	0	0.6	0	0.4
4	0	0	0	0	1.0

To compute p^2 one row at a time we note:

$p^2(0,0) = 1$ and $p^2(4,4) = 1$, since these are absorbing states.
$p^2(1,3) = (.4)^2 = 0.16$, since the chain has to go up twice.
$p^2(1,1) = (.4)(.6) = 0.24$. The chain must go from 1 to 2 to 1.
$p^2(1,0) = 0.6$. To be at 0 at time 2, the first jump must be to 0.

Leaving the cases $i = 2, 3$ to the reader, we have

$$p^2 = \begin{pmatrix} 1.0 & 0 & 0 & 0 & 0 \\ .6 & .24 & 0 & .16 & 0 \\ .36 & 0 & .48 & 0 & .16 \\ 0 & .36 & 0 & .24 & .4 \\ 0 & 0 & 0 & 0 & 1 \end{pmatrix}$$

Using a calculator one can easily compute

$$p^{20} = \begin{pmatrix} 1.0 & 0 & 0 & 0 & 0 \\ .87655 & .00032 & 0 & .00022 & .12291 \\ .69186 & 0 & .00065 & 0 & .30749 \\ .41842 & .00049 & 0 & .00032 & .58437 \\ 0 & 0 & 0 & 0 & 1 \end{pmatrix}$$

0 and 4 are absorbing states. Here we see that the probability of avoiding absorption for 20 steps is 0.00054 from state 1, 0.00065 from state 2, and 0.00081 from state 3.

Later we will see that

$$\lim_{n \to \infty} p^n = \begin{pmatrix} 1.0 & 0 & 0 & 0 & 0 \\ 57/65 & 0 & 0 & 0 & 8/65 \\ 45/65 & 0 & 0 & 0 & 20/65 \\ 27/65 & 0 & 0 & 0 & 38/65 \\ 0 & 0 & 0 & 0 & 1 \end{pmatrix}$$

1.3   Classification of States

We begin with some important notation. We are often interested in the behavior of the chain for a fixed initial state, so we will introduce the shorthand

$$P_x(A) = P(A|X_0 = x)$$

Later we will have to consider expected values for this probability and we will denote them by E_x.

Let $T_y = \min\{n \geq 1 : X_n = y\}$ be the **time of the first return to** y (i.e., being there at time 0 doesn't count), and let

$$\rho_{yy} = P_y(T_y < \infty)$$

be the probability X_n returns to y when it starts at y. Note that if we didn't exclude $n = 0$ this probability would always be 1.

Intuitively, the Markov property implies that the probability X_n will return to y at least twice is ρ_{yy}^2, since after the first return, the chain is at y, and the probability of a second return following the first is again ρ_{yy}.

To show that the reasoning in the last paragraph is valid, we have to introduce a definition and state a theorem. We say that T is a **stopping time** if the occurrence (or nonoccurrence) of the event "we stop at time n," $\{T = n\}$, can be determined by looking at the values of the process up to that time: $X_0, \ldots, X_n$. To see that T_y is a stopping time note that

$$\{T_y = n\} = \{X_1 \neq y, \ldots, X_{n-1} \neq y, X_n = y\}$$

and that the right-hand side can be determined from $X_0, \ldots, X_n$.

Since stopping at time n depends only on the values $X_0, \ldots, X_n$, and in a Markov chain the distribution of the future only depends on the past through the current state, it should not be hard to believe that the Markov property holds at stopping times. This fact can be stated formally as

Theorem 1.2 (Strong Markov Property). *Suppose T is a stopping time. Given that $T = n$ and $X_T = y$, any other information about $X_0, \ldots X_T$ is irrelevant for*

predicting the future, and X_{T+k}, $k \geq 0$ behaves like the Markov chain with initial state y.

Why is this true? To keep things as simple as possible we will show only that

$$P(X_{T+1} = z | X_T = y, T = n) = p(y, z)$$

Let V_n be the set of vectors $(x_0, \ldots, x_n)$ so that if $X_0 = x_0, \ldots, X_n = x_n$, then $T = n$ and $X_T = y$. Breaking things down according to the values of $X_0, \ldots, X_n$ gives

$$P(X_{T+1} = z, X_T = y, T = n) = \sum_{x \in V_n} P(X_{n+1} = z, X_n = x_n, \ldots, X_0 = x_0)$$

$$= \sum_{x \in V_n} P(X_{n+1} = z | X_n = x_n, \ldots, X_0 = x_0) P(X_n = x_n, \ldots, X_0 = x_0)$$

where in the second step we have used the **multiplication rule**

$$P(A \cap B) = P(B|A)P(A)$$

For any $(x_0, \ldots, x_n) \in V_n$ we have $T = n$ and $X_T = y$ so $x_n = y$. Using the Markov property, (1.1), and recalling the definition of V_n shows the above

$$P(X_{T+1} = z, T = n, X_T = y) = p(y, z) \sum_{x \in V_n} P(X_n = x_n, \ldots, X_0 = x_0)$$

$$= p(y, z)P(T = n, X_T = y)$$

Dividing both sides by $P(T = n, X_T = y)$ gives the desired result.   □

Let $T_y^1 = T_y$ and for $k \geq 2$ let

$$T_y^k = \min\{n > T_y^{k-1} : X_n = y\} \tag{1.3}$$

be the **time of the kth return to** y. The strong Markov property implies that the conditional probability we will return one more time given that we have returned $k - 1$ times is ρ_{yy}. This and induction imply that

$$P_y(T_y^k < \infty) = \rho_{yy}^k \tag{1.4}$$

At this point, there are two possibilities:

(i) $\rho_{yy} < 1$: The probability of returning k times is $\rho_{yy}^k \to 0$ as $k \to \infty$. Thus, eventually the Markov chain does not find its way back to y. In this case the state y is called **transient**, since after some point it is never visited by the Markov chain.

(ii) $\rho_{yy} = 1$: The probability of returning k times $\rho_{yy}^k = 1$, so the chain returns to y infinitely many times. In this case, the state y is called **recurrent**, it continually recurs in the Markov chain.

To understand these notions, we turn to our examples, beginning with

Example 1.12 (Gambler's Ruin). Consider, for concreteness, the case $N = 4$.

$$
\begin{array}{c|ccccc}
 & 0 & 1 & 2 & 3 & 4 \\
\hline
0 & 1 & 0 & 0 & 0 & 0 \\
1 & .6 & 0 & .4 & 0 & 0 \\
2 & 0 & .6 & 0 & .4 & 0 \\
3 & 0 & 0 & .6 & 0 & .4 \\
4 & 0 & 0 & 0 & 0 & 1 \\
\end{array}
$$

We will show that eventually the chain gets stuck in either the bankrupt (0) or happy winner (4) state. In the terms of our recent definitions, we will show that states $0 < y < 4$ are transient, while the states 0 and 4 are recurrent.

It is easy to check that 0 and 4 are recurrent. Since $p(0,0) = 1$, the chain comes back on the next step with probability one, i.e.,

$$P_0(T_0 = 1) = 1$$

and hence $\rho_{00} = 1$. A similar argument shows that 4 is recurrent. In general if y is an absorbing state, i.e., if $p(y, y) = 1$, then y is a very strongly recurrent state—the chain always stays there.

To check the transience of the interior states, $1, 2, 3$, we note that starting from 1, if the chain goes to 0, it will never return to 1, so the probability of never returning to 1,

$$P_1(T_1 = \infty) \geq p(1, 0) = 0.6 > 0$$

Similarly, starting from 2, the chain can go to 1 and then to 0, so

$$P_2(T_2 = \infty) \geq p(2, 1)p(1, 0) = 0.36 > 0$$

Finally, for starting from 3, we note that the chain can go immediately to 4 and never return with probability 0.4, so

$$P_3(T_3 = \infty) \geq p(3, 4) = 0.4 > 0$$

In some cases it is easy to identify recurrent states.

Example 1.13 (Social Mobility). Recall that the transition probability is

$$
\begin{array}{c c c c}
 & \mathbf{1} & \mathbf{2} & \mathbf{3} \\
\mathbf{1} & .7 & .2 & .1 \\
\mathbf{2} & .3 & .5 & .2 \\
\mathbf{3} & .2 & .4 & .4
\end{array}
$$

To begin we note that no matter where X_n is, there is a probability of at least 0.1 of hitting 3 on the next step so

$$P_3(T_3 > n) \le (0.9)^n \to 0 \text{ as } n \to \infty$$

i.e., we will return to 3 with probability 1. The last argument applies even more strongly to states 1 and 2, since the probability of jumping to them on the next step is always at least 0.2. Thus all three states are recurrent.

The last argument generalizes to give the following useful fact.

Lemma 1.3. *Suppose $P_x(T_y \le k) \ge \alpha > 0$ for all x in the state space S. Then*

$$P_x(T_y > nk) \le (1 - \alpha)^n$$

Generalizing from our experience with the last two examples, we will introduce some general results that will help us identify transient and recurrent states.

Definition 1.1. We say that x **communicates with** y and write $x \to y$ if there is a positive probability of reaching y starting from x, that is, the probability

$$\rho_{xy} = P_x(T_y < \infty) > 0$$

Note that the last probability includes not only the possibility of jumping from x to y in one step but also going from x to y after visiting several other states in between. The following property is simple but useful. Here and in what follows, lemmas are a means to prove the more important conclusions called theorems. The two are numbered in the same sequence to make results easier to find.

Lemma 1.4. *If $x \to y$ and $y \to z$, then $x \to z$.*

Proof. Since $x \to y$ there is an m so that $p^m(x, y) > 0$. Similarly there is an n so that $p^n(y, z) > 0$. Since $p^{m+n}(x, z) \ge p^m(x, y)p^n(y, z)$ it follows that $x \to z$.   □

Theorem 1.5. *If $\rho_{xy} > 0$, but $\rho_{yx} < 1$, then x is transient.*

Proof. Let $K = \min\{k : p^k(x, y) > 0\}$ be the smallest number of steps we can take to get from x to y. Since $p^K(x, y) > 0$ there must be a sequence $y_1, \ldots y_{K-1}$ so that

$$p(x, y_1)p(y_1, y_2) \cdots p(y_{K-1}, y) > 0$$

Since K is minimal all the $y_i \neq x$ (or there would be a shorter path), and we have

$$P_x(T_x = \infty) \geq p(x, y_1)p(y_1, y_2) \cdots p(y_{K-1}, y)(1 - \rho_{yx}) > 0$$

so x is transient.   □

We will see later that Theorem 1.5 allows us to identify all the transient states when the state space is finite. An immediate consequence of Theorem 1.5 is

Lemma 1.6. *If x is recurrent and $\rho_{xy} > 0$, then $\rho_{yx} = 1$.*

Proof. If $\rho_{yx} < 1$, then Lemma 1.5 would imply x is transient.   □

To be able to analyze any finite state Markov chain we need some theory. To motivate the developments consider

Example 1.14 (A Seven-State Chain). Consider the transition probability:

```
      1  2  3  4  5  6  7
1   .7  0  0  0 .3  0  0
2   .1 .2 .3 .4  0  0  0
3    0  0 .5 .3 .2  0  0
4    0  0  0 .5  0 .5  0
5   .6  0  0  0 .4  0  0
6    0  0  0  0  0 .2 .8
7    0  0  0  1  0  0  0
```

To identify the states that are recurrent and those that are transient, we begin by drawing a graph that will contain an arc from i to j if $p(i,j) > 0$ and $i \neq j$. We do not worry about drawing the self-loops corresponding to states with $p(i,i) > 0$ since such transitions cannot help the chain get somewhere new.

In the case under consideration the graph is

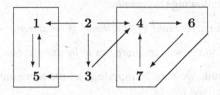

The state 2 communicates with 1, which does not communicate with it, so Theorem 1.5 implies that 2 is transient. Likewise 3 communicates with 4, which doesn't communicate with it, so 3 is transient. To conclude that all the remaining states are recurrent we will introduce two definitions and a fact.

A set A is **closed** if it is impossible to get out, i.e., if $i \in A$ and $j \notin A$ then $p(i,j) = 0$. In Example 1.14, $\{1,5\}$ and $\{4,6,7\}$ are closed sets. Their union

$\{1, 4, 5, 6, 7\}$ is also closed. One can add 3 to get another closed set $\{1, 3, 4, 5, 6, 7\}$. Finally, the whole state space $\{1, 2, 3, 4, 5, 6, 7\}$ is always a closed set.

Among the closed sets in the last example, some are obviously too big. To rule them out, we need a definition. A set B is called **irreducible** if whenever $i, j \in B$, i communicates with j. The irreducible closed sets in the Example 1.14 are $\{1, 5\}$ and $\{4, 6, 7\}$. The next result explains our interest in irreducible closed sets.

Theorem 1.7. *If C is a finite closed and irreducible set, then all states in C are recurrent.*

Before entering into an explanation of this result, we note that Theorem 1.7 tells us that 1, 5, 4, 6, and 7 are recurrent, completing our study of the Example 1.14 with the results we had claimed earlier.

In fact, the combination of Theorems 1.5 and 1.7 is sufficient to classify the states in any finite state Markov chain. An algorithm will be explained in the proof of the following result.

Theorem 1.8. *If the state space S is finite, then S can be written as a disjoint union $T \cup R_1 \cup \cdots \cup R_k$, where T is a set of transient states and the R_i, $1 \leq i \leq k$, are closed irreducible sets of recurrent states.*

Proof. Let T be the set of x for which there is a y so that $x \to y$ but $y \not\to x$. The states in T are transient by Theorem 1.5. Our next step is to show that all the remaining states, $S - T$, are recurrent.

Pick an $x \in S - T$ and let $C_x = \{y : x \to y\}$. Since $x \notin T$ it has the property if $x \to y$, then $y \to x$. To check that C_x is closed note that if $y \in C_x$ and $y \to z$, then Lemma 1.4 implies $x \to z$ so $z \in C_x$. To check irreducibility, note that if $y, z \in C_x$, then by our first observation $y \to x$ and we have $x \to z$ by definition, so Lemma 1.4 implies $y \to z$. C_x is closed and irreducible so all states in C_x are recurrent. Let $R_1 = C_x$. If $S - T - R_1 = \emptyset$, we are done. If not, pick a site $w \in S - T - R_1$ and repeat the procedure. $\quad\square$

The rest of this section is devoted to the proof of Theorem 1.7. To do this, it is enough to prove the following two results.

Lemma 1.9. *If x is recurrent and $x \to y$, then y is recurrent.*

Lemma 1.10. *In a finite closed set there has to be at least one recurrent state.*

To prove these results we need to introduce a little more theory. Recall the time of the kth visit to y defined by

$$T_y^k = \min\{n > T_y^{k-1} : X_n = y\}$$

and $\rho_{xy} = P_x(T_y < \infty)$ the probability we ever visit y at some time $n \geq 1$ when we start from x. Using the strong Markov property as in the proof of (1.4) gives

$$P_x(T_y^k < \infty) = \rho_{xy}\rho_{yy}^{k-1}. \qquad (1.5)$$

Let $N(y)$ be the number of visits to y at times $n \geq 1$. Using (1.5) we can compute $EN(y)$.

Lemma 1.11. $E_x N(y) = \rho_{xy}/(1 - \rho_{yy})$

Proof. Accept for the moment the fact that for any nonnegative integer valued random variable X, the expected value of X can be computed by

$$EX = \sum_{k=1}^{\infty} P(X \ge k) \tag{1.6}$$

We will prove this after we complete the proof of Lemma 1.11. Now the probability of returning at least k times, $\{N(y) \ge k\}$, is the same as the event that the kth return occurs, i.e., $\{T_y^k < \infty\}$, so using (1.5) we have

$$E_x N(y) = \sum_{k=1}^{\infty} P(N(y) \ge k) = \rho_{xy} \sum_{k=1}^{\infty} \rho_{yy}^{k-1} = \frac{\rho_{xy}}{1 - \rho_{yy}}$$

since $\sum_{n=0}^{\infty} \theta^n = 1/(1 - \theta)$ whenever $|\theta| < 1$. □

Proof of (1.6). Let $1_{\{X \ge k\}}$ denote the random variable that is 1 if $X \ge k$ and 0 otherwise. It is easy to see that

$$X = \sum_{k=1}^{\infty} 1_{\{X \ge k\}}.$$

Taking expected values and noticing $E1_{\{X \ge k\}} = P(X \ge k)$ gives

$$EX = \sum_{k=1}^{\infty} P(X \ge k)$$

□

Our next step is to compute the expected number of returns to y in a different way.

Lemma 1.12. $E_x N(y) = \sum_{n=1}^{\infty} p^n(x, y)$.

Proof. Let $1_{\{X_n = y\}}$ denote the random variable that is 1 if $X_n = y$, 0 otherwise. Clearly

$$N(y) = \sum_{n=1}^{\infty} 1_{\{X_n = y\}}.$$

Taking expected values now gives

$$E_x N(y) = \sum_{n=1}^{\infty} P_x(X_n = y)$$

□

With the two lemmas established we can now state our next main result.

Theorem 1.13. *y is recurrent if and only if*

$$\sum_{n=1}^{\infty} p^n(y, y) = E_y N(y) = \infty$$

Proof. The first equality is Lemma 1.12. From Lemma 1.11 we see that $E_y N(y) = \infty$ if and only if $\rho_{yy} = 1$, which is the definition of recurrence.   □

With this established we can easily complete the proofs of our two lemmas .

Proof of Lemma 1.9. Suppose x is recurrent and $\rho_{xy} > 0$. By Lemma 1.6 we must have $\rho_{yx} > 0$. Pick j and ℓ so that $p^j(y, x) > 0$ and $p^\ell(x, y) > 0$. $p^{j+k+\ell}(y, y)$ is probability of going from y to y in $j + k + \ell$ steps while the product $p^j(y, x) p^k(x, x) p^\ell(x, y)$ is the probability of doing this and being at x at times j and $j + k$. Thus we must have

$$\sum_{k=0}^{\infty} p^{j+k+\ell}(y, y) \geq p^j(y, x) \left(\sum_{k=0}^{\infty} p^k(x, x) \right) p^\ell(x, y)$$

If x is recurrent, then $\sum_k p^k(x, x) = \infty$, so $\sum_m p^m(y, y) = \infty$ and Theorem 1.13 implies that y is recurrent.   □

Proof of Lemma 1.10. If all the states in C are transient, then Lemma 1.11 implies that $E_x N(y) < \infty$ for all x and y in C. Since C is finite, using Lemma 1.12

$$\infty > \sum_{y \in C} E_x N(y) = \sum_{y \in C} \sum_{n=1}^{\infty} p^n(x, y)$$

$$= \sum_{n=1}^{\infty} \sum_{y \in C} p^n(x, y) = \sum_{n=1}^{\infty} 1 = \infty$$

where in the next to last equality we have used that C is closed. This contradiction proves the desired result.   □

1.4 Stationary Distributions

In the next section we will see that if we impose an additional assumption called aperiodicity an irreducible finite state Markov chain converges to a stationary distribution

$$p^n(x, y) \to \pi(y)$$

To prepare for that this section introduces stationary distributions and shows how to compute them. Our first step is to consider
What happens in a Markov chain when the initial state is random? Breaking things down according to the value of the initial state and using the definition of conditional probability

$$P(X_n = j) = \sum_i P(X_0 = i, X_n = j)$$

$$= \sum_i P(X_0 = i)P(X_n = j|X_0 = i)$$

If we introduce $q(i) = P(X_0 = i)$, then the last equation can be written as

$$P(X_n = j) = \sum_i q(i)p^n(i, j) \tag{1.7}$$

In words, we multiply the transition matrix on the left by the vector q of initial probabilities. If there are k states, then $p^n(x, y)$ is a $k \times k$ matrix. So to make the matrix multiplication work out right, we should take q as a $1 \times k$ matrix or a "row vector."

Example 1.15. Consider the weather chain (Example 1.3) and suppose that the initial distribution is $q(1) = 0.3$ and $q(2) = 0.7$. In this case

$$(.3 \ .7) \begin{pmatrix} .6 & .4 \\ .2 & .8 \end{pmatrix} = (.32 \ .68)$$

since $.3(.6) + .7(.2) = .32$

$.3(.4) + .7(.8) = .68$

Example 1.16. Consider the social mobility chain (Example 1.4) and suppose that the initial distribution: $q(1) = .5$, $q(2) = .2$, and $q(3) = .3$. Multiplying the vector q by the transition probability gives the vector of probabilities at time 1.

$$(.5 \ .2 \ .3) \begin{pmatrix} .7 & .2 & .1 \\ .3 & .5 & .2 \\ .2 & .4 & .4 \end{pmatrix} = (.47 \ .32 \ .21)$$

To check the arithmetic note that the three entries on the right-hand side are

$$.5(.7) + .2(.3) + .3(.2) = .35 + .06 + .06 = .47$$
$$.5(.2) + .2(.5) + .3(.4) = .10 + .10 + .12 = .32$$
$$.5(.1) + .2(.2) + .3(.4) = .05 + .04 + .12 = .21$$

If $qp = q$, then q is called a **stationary distribution**. If the distribution at time 0 is the same as the distribution at time 1, then by the Markov property it will be the distribution at all times $n \geq 1$.

Stationary distributions have a special importance in the theory of Markov chains, so we will use a special letter π to denote solutions of the equation

$$\pi p = \pi.$$

To have a mental picture of what happens to the distribution of probability when one step of the Markov chain is taken, it is useful to think that we have $q(i)$ pounds of sand at state i, with the total amount of sand $\sum_i q(i)$ being one pound. When a step is taken in the Markov chain, a fraction $p(i,j)$ of the sand at i is moved to j. The distribution of sand when this has been done is

$$qp = \sum_i q(i)p(i,j)$$

If the distribution of sand is not changed by this procedure q is a stationary distribution.

Example 1.17 (Weather Chain). To compute the stationary distribution we want to solve

$$(\pi_1 \ \pi_2) \begin{pmatrix} .6 & .4 \\ .2 & .8 \end{pmatrix} = (\pi_1 \ \pi_2)$$

Multiplying gives two equations:

$$.6\pi_1 + .2\pi_2 = \pi_1$$
$$.4\pi_1 + .8\pi_2 = \pi_2$$

Both equations reduce to $.4\pi_1 = .2\pi_2$. Since we want $\pi_1 + \pi_2 = 1$, we must have $.4\pi_1 = .2 - .2\pi_1$, and hence

$$\pi_1 = \frac{.2}{.2 + .4} = \frac{1}{3} \qquad \pi_2 = \frac{.4}{.2 + .4} = \frac{2}{3}$$

To check this we note that

$$(1/3 \ 2/3) \begin{pmatrix} .6 & .4 \\ .2 & .8 \end{pmatrix} = \left(\frac{.6}{3} + \frac{.4}{3}, \frac{.4}{3} + \frac{1.6}{3} \right)$$

General two state transition probability.

$$
\begin{array}{ccc}
 & 1 & 2 \\
1 & 1-a & a \\
2 & b & 1-b
\end{array}
$$

We have written the chain in this way so the stationary distribution has a simple formula

$$\pi_1 = \frac{b}{a+b} \qquad \pi_2 = \frac{a}{a+b} \tag{1.8}$$

As a first check on this formula we note that in the weather chain $a = 0.4$ and $b = 0.2$ which gives $(1/3, 2/3)$ as we found before. We can prove this works in general by drawing a picture:

$$
\frac{b}{a+b} \quad \overset{\displaystyle 1 \quad a}{\underset{b}{\rightleftharpoons}} \quad \overset{\displaystyle 2}{\bullet} \quad \frac{a}{a+b}
$$

In words, the amount of sand that flows from 1 to 2 is the same as the amount that flows from 2 to 1 so the amount of sand at each site stays constant. To check algebraically that $\pi p = \pi$:

$$\frac{b}{a+b}(1-a) + \frac{a}{a+b}b = \frac{b-ba+ab}{a+b} = \frac{b}{a+b}$$

$$\frac{b}{a+b}a + \frac{a}{a+b}(1-b) = \frac{ba+a-ab}{a+b} = \frac{a}{a+b} \tag{1.9}$$

Formula (1.8) gives the stationary distribution for any two state chain, so we progress now to the three state case and consider the

Example 1.18 (Social Mobility (Continuation of 1.4)).

$$
\begin{array}{cccc}
 & 1 & 2 & 3 \\
1 & .7 & .2 & .1 \\
2 & .3 & .5 & .2 \\
3 & .2 & .4 & .4
\end{array}
$$

The equation $\pi p = \pi$ says

$$
\begin{pmatrix} \pi_1 & \pi_2 & \pi_3 \end{pmatrix}
\begin{pmatrix} .7 & .2 & .1 \\ .3 & .5 & .2 \\ .2 & .4 & .4 \end{pmatrix}
= \begin{pmatrix} \pi_1 & \pi_2 & \pi_3 \end{pmatrix}
$$

which translates into three equations

$$.7\pi_1 + .3\pi_2 + .2\pi_3 = \pi_1$$
$$.2\pi_1 + .5\pi_2 + .4\pi_3 = \pi_2$$
$$.1\pi_1 + .2\pi_2 + .4\pi_3 = \pi_3$$

Note that the columns of the matrix give the numbers in the rows of the equations. The third equation is redundant since if we add up the three equations we get

$$\pi_1 + \pi_2 + \pi_3 = \pi_1 + \pi_2 + \pi_3$$

If we replace the third equation by $\pi_1 + \pi_2 + \pi_3 = 1$ and subtract π_1 from each side of the first equation and π_2 from each side of the second equation we get

$$-.3\pi_1 + .3\pi_2 + .2\pi_3 = 0$$
$$.2\pi_1 - .5\pi_2 + .4\pi_3 = 0$$
$$\pi_1 + \pi_2 + \pi_3 = 1 \qquad (1.10)$$

At this point we can solve the equations by hand or using a calculator.

By Hand We note that the third equation implies $\pi_3 = 1 - \pi_1 - \pi_2$ and substituting this in the first two gives

$$.2 = .5\pi_1 - .1\pi_2$$
$$.4 = .2\pi_1 + .9\pi_2$$

Multiplying the first equation by .9 and adding .1 times the second gives

$$2.2 = (0.45 + 0.02)\pi_1 \qquad \text{or} \qquad \pi_1 = 22/47$$

Multiplying the first equation by .2 and adding $-.5$ times the second gives

$$-0.16 = (-.02 - 0.45)\pi_2 \qquad \text{or} \qquad \pi_2 = 16/47$$

Since the three probabilities add up to 1, $\pi_3 = 9/47$.

Using the TI83 calculator is easier. To begin we write (1.10) in matrix form as

$$(\pi_1 \ \pi_2 \ \pi_3) \begin{pmatrix} -.3 & .2 & 1 \\ .3 & -.5 & 1 \\ .2 & .4 & 1 \end{pmatrix} = (0 \ 0 \ 1)$$

If we let A be the 3×3 matrix in the middle this can be written as $\pi A = (0, 0, 1)$. Multiplying on each side by A^{-1} we see that

$$\pi = (0, 0, 1)A^{-1}$$

which is the third row of A^{-1}. To compute A^{-1}, we enter A into our calculator (using the MATRX menu and its EDIT submenu), use the MATRIX menu to put $[A]$ on the computation line, press x^{-1}, and then ENTER. Reading the third row we find that the stationary distribution is

$$(0.468085, 0.340425, 0.191489)$$

Converting the answer to fractions using the first entry in the MATH menu gives

$$(22/47, 16/47, 9/47)$$

Example 1.19 (Brand Preference (Continuation of 1.5)).

$$
\begin{array}{c c c c}
 & \mathbf{1} & \mathbf{2} & \mathbf{3} \\
\mathbf{1} & .8 & .1 & .1 \\
\mathbf{2} & .2 & .6 & .2 \\
\mathbf{3} & .3 & .3 & .4 \\
\end{array}
$$

Using the first two equations and the fact that the sum of the π's is 1

$$.8\pi_1 + .2\pi_2 + .3\pi_3 = \pi_1$$
$$.1\pi_1 + .6\pi_2 + .3\pi_3 = \pi_2$$
$$\pi_1 + \pi_2 + \pi_3 = 1$$

Subtracting π_1 from both sides of the first equation and π_2 from both sides of the second, this translates into $\pi A = (0, 0, 1)$ with

$$A = \begin{pmatrix} -.2 & .1 & 1 \\ .2 & -.4 & 1 \\ .3 & .3 & 1 \end{pmatrix}$$

Note that here and in the previous example the first two columns of A consist of the first two columns of the transition probability with 1 subtracted from the diagonal entries, and the final column is all 1's. Computing the inverse and reading the last row gives

$$(0.545454, 0.272727, 0.181818)$$

Converting the answer to fractions using the first entry in the MATH menu gives

$$(6/11, 3/11, 2/11)$$

To check this we note that

$$\left(6/11\ 3/11\ 2/11\right)\begin{pmatrix}.8 & .1 & .1 \\ .2 & .6 & .2 \\ .3 & .3 & .4\end{pmatrix}$$

$$=\left(\frac{4.8+.6+.6}{11}\ \frac{.6+1.8+.6}{11}\ \frac{.6+.6+.8}{11}\right)$$

Example 1.20 (Hot Hand (Continuation of 1.10)). To find the stationary matrix in this case we can follow the same procedure. A consists of the first three columns of the transition matrix with 1 subtracted from the diagonal, and a final column of all 1's.

$$\begin{matrix} -1/4 & 1/4 & 0 & 1 \\ 0 & -1 & 2/3 & 1 \\ 2/3 & 1/3 & -1 & 1 \\ 0 & 0 & 1/2 & 1 \end{matrix}$$

The answer is given by the fourth row of A^{-1}:

$$(0.5,\ 0.1875,\ 0.1875,\ 0.125) = (1/2,\ 3/16,\ 3/16,\ 1/8)$$

Thus the long run fraction of time the player hits a shot is

$$\pi(HH) + \pi(MH) = 0.6875 = 11/36.$$

1.4.1 Doubly Stochastic Chains

Definition 1.2. A transition matrix p is said to be **doubly stochastic** if its COLUMNS sum to 1, or in symbols $\sum_x p(x, y) = 1$.

The adjective "doubly" refers to the fact that by its definition a transition probability matrix has ROWS that sum to 1, i.e., $\sum_y p(x, y) = 1$. The stationary distribution is easy to guess in this case:

Theorem 1.14. *If p is a doubly stochastic transition probability for a Markov chain with N states, then the uniform distribution, $\pi(x) = 1/N$ for all x, is a stationary distribution.*

Proof. To check this claim we note that if $\pi(x) = 1/N$ then

$$\sum_x \pi(x)p(x, y) = \frac{1}{N}\sum_x p(x, y) = \frac{1}{N} = \pi(y)$$

Looking at the second equality we see that conversely, if $\pi(x) = 1/N$ then p is doubly stochastic.   □

Example 1.21 (Symmetric Reflecting Random Walk on the Line). The state space is $\{0, 1, 2 \ldots, L\}$. The chain goes to the right or left at each step with probability 1/2, subject to the rules that if it tries to go to the left from 0 or to the right from L it stays put. For example, when $L = 4$ the transition probability is

	0	1	2	3	4
0	0.5	0.5	0	0	0
1	0.5	0	0.5	0	0
2	0	0.5	0	0.5	0
3	0	0	0.5	0	0.5
4	0	0	0	0.5	0.5

It is clear in the example $L = 4$ that each column adds up to 1. With a little thought one sees that this is true for any L, so the stationary distribution is uniform, $\pi(i) = 1/(L+1)$.

Example 1.22 (Tiny Board Game). Consider a circular board game with only six spaces $\{0, 1, 2, 3, 4, 5\}$. On each turn we roll a die with 1 on three sides, 2 on two sides, and 3 on one side to decide how far to move. Here we consider 5 to be adjacent to 0, so if we are there and we roll a 2 then the result is $5 + 2 \bmod 6 = 1$, where $i + k \bmod 6$ is the remainder when $i + k$ is divided by 6. In this case the transition probability is

	0	1	2	3	4	5
0	0	1/3	1/3	1/6	0	0
1	0	0	1/2	1/3	1/6	0
2	0	0	0	1/2	1/3	1/6
3	1/6	0	0	0	1/2	1/3
4	1/3	1/6	0	0	0	1/2
5	1/2	1/3	1/6	0	0	0

It is clear that the columns add to one, so the stationary distribution is uniform. To check the hypothesis of the convergence theorem, we note that after 3 turns we will have moved between three and nine spaces so $p^3(i, j) > 0$ for all i and j.

Example 1.23 (Mathematician's Monopoly). The game Monopoly is played on a game board that has 40 spaces arranged around the outside of a square. The squares have names like *Reading Railroad* and *Park Place* but we will number the squares 0 (*Go*), 1 (*Baltic Avenue*), ... 39 (*Boardwalk*). In Monopoly you roll two dice and move forward a number of spaces equal to the sum. For the moment, we will ignore things like *Go to Jail, Chance,* and other squares that make the game more interesting and formulate the dynamics as following. Let r_k be the probability that the sum of two dice is k ($r_2 = 1/36$, $r_3 = 2/36$, ... $r_7 = 6/36$, ..., $r_{12} = 1/36$) and let

$$p(i,j) = r_k \qquad \text{if } j = i + k \bmod 40$$

where $i+k \bmod 40$ is the remainder when $i+k$ is divided by 40. To explain suppose that we are sitting on *Park Place* $i = 37$ and roll $k = 6$. $37 + 6 = 43$ but when we divide by 40 the remainder is 3, so $p(37,3) = r_6 = 5/36$.

This example is larger but has the same structure as the previous example. Each row has the same entries but shifts one unit to the right each time with the number that goes off the right edge emerging in the 0 column. This structure implies that each entry in the row appears once in each column and hence the sum of the entries in the column is 1, and the stationary distribution is uniform. To check the hypothesis of the convergence theorem note that in four rolls you can move forward by 8–48 squares, so $p^4(i,j) > 0$ for all i and j.

Example 1.24. **Real Monopoly** has two complications:

- Square 30 is "Go to Jail," which sends you to square 10. You can buy your way out of jail but in the results we report below, we assume that you are cheap. If you roll a double, then you get out for free. If you don't get doubles in three tries, you have to pay.
- There are three *Chance* squares at 7, 12, and 36 (diamonds on the graph), and three *Community Chest* squares at 2, 17, 33 (squares on the graph), where you draw a card, which can send you to another square.

The graph gives the long run frequencies of being in different squares on the Monopoly board at the end of your turn, as computed by simulation (Fig. 1.1). We have removed the 9.46 % chance of being *In Jail* to make the probabilities easier to see. The value reported for square 10 is the 2.14 % probability of *Just Visiting Jail*, i.e., being brought there by the roll of the dice. Square 30, *Go to Jail*, has probability 0 for the obvious reasons. The other three lowest values occur for *Chance* squares. Due to the transition from 30 to 10, frequencies for squares near 20 are increased relative to the average of 2.5 % while those after 30 or before 10 are decreased. Squares 0 (*Go*) and 5 (*Reading Railroad*) are exceptions to this trend since there are Chance cards that instruct you to go there.

1.5   Detailed Balance Condition

π is said to satisfy the **detailed balance condition** if

$$\pi(x)p(x,y) = \pi(y)p(y,x) \tag{1.11}$$

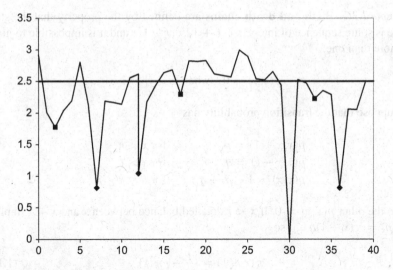

Fig. 1.1 Stationary distribution for monopoly

To see that this is a stronger condition than $\pi p = \pi$, we sum over x on each side to get

$$\sum_x \pi(x)p(x,y) = \pi(y) \sum_x p(y,x) = \pi(y)$$

As in our earlier discussion of stationary distributions, we think of $\pi(x)$ as giving the amount of sand at x, and one transition of the chain as sending a fraction $p(x,y)$ of the sand at x to y. In this case the detailed balance condition says that the amount of sand going from x to y in one step is exactly balanced by the amount going back from y to x. In contrast the condition $\pi p = \pi$ says that after all the transfers are made, the amount of sand that ends up at each site is the same as the amount that starts there.

Many chains do not have stationary distributions that satisfy the detailed balance condition.

Example 1.25. Consider

$$
\begin{array}{c|ccc}
 & 1 & 2 & 3 \\
\hline
1 & .5 & .5 & 0 \\
2 & .3 & .1 & .6 \\
3 & .2 & .4 & .4 \\
\end{array}
$$

There is no stationary distribution with detailed balance since $\pi(1)p(1,3) = 0$ but $p(1,3) > 0$ so we would have to have $\pi(3) = 0$ and using $\pi(3)p(3,i) = \pi(i)p(i,3)$ we conclude all the $\pi(i) = 0$. This chain is doubly stochastic so $(1/3, 1/3, 1/3)$ is a stationary distribution.

Example 1.26. **Birth and death chains** are defined by the property that the state space is some sequence of integers $\ell, \ell + 1, \ldots r - 1, r$ and it is impossible to jump by more than one:

$$p(x, y) = 0 \quad \text{when } |x - y| > 1$$

Suppose that the transition probability has

$$
\begin{aligned}
p(x, x + 1) &= p_x && \text{for } x < r \\
p(x, x - 1) &= q_x && \text{for } x > \ell \\
p(x, x) &= 1 - p_x - q_x && \text{for } \ell \le x \le r
\end{aligned}
$$

while the other $p(x, y) = 0$. If $x < r$ detailed balance between x and $x + 1$ implies $\pi(x)p_x = \pi(x + 1)q_{x+1}$, so

$$\pi(x + 1) = \frac{p_x}{q_{x+1}} \cdot \pi(x) \tag{1.12}$$

Using this with $x = \ell$ gives $\pi(\ell + 1) = \pi(\ell)p_\ell/q_{\ell+1}$. Taking $x = \ell + 1$

$$\pi(\ell + 2) = \frac{p_{\ell+1}}{q_{\ell+2}} \cdot \pi(\ell + 1) = \frac{p_{\ell+1} \cdot p_\ell}{q_{\ell+2} \cdot q_{\ell+1}} \cdot \pi(\ell)$$

Extrapolating from the first two results we see that in general

$$\pi(\ell + i) = \pi(\ell) \cdot \frac{p_{\ell+i-1} \cdot p_{\ell+i-2} \cdots p_{\ell+1} \cdot p_\ell}{q_{\ell+i} \cdot q_{\ell+i-1} \cdots q_{\ell+2} \cdot q_{\ell+1}}$$

To keep the indexing straight note that: (i) there are i terms in the numerator and in the denominator, (ii) the indices decrease by 1 each time, (iii) the answer will not depend on q_ℓ (which is 0) or $p_{\ell+i}$.

For a concrete example to illustrate the use of this formula consider

Example 1.27 (Ehrenfest Chain). For concreteness, suppose there are three balls. In this case the transition probability is

	0	1	2	3
0	0	3/3	0	0
1	1/3	0	2/3	0
2	0	2/3	0	1/3
3	0	0	3/3	0

Setting $\pi(0) = c$ and using (1.12) we have

$$\pi(1) = 3c, \qquad \pi(2) = \pi(1) = 3c \qquad \pi(3) = \pi(2)/3 = c.$$

The sum of the π's is $8c$, so we pick $c = 1/8$ to get

$$\pi(0) = 1/8, \qquad \pi(1) = 3/8, \qquad \pi(2) = 3/8, \qquad \pi(3) = 1/8$$

Knowing the answer, one can look at the last equation and see that π represents the distribution of the number of Heads when we flip three coins, then guess and verify that in general that the binomial distribution with $p = 1/2$ is the stationary distribution:

$$\pi(x) = 2^{-n}\binom{n}{x}$$

Here $m! = 1 \cdot 2 \cdots (m-1) \cdot m$, with $0! = 1$, and

$$\binom{n}{x} = \frac{n!}{x!(n-x)!}$$

is the binomial coefficient which gives the number of ways of choosing x objects out of a set of n.

To check that our guess satisfies the detailed balance condition, we note that

$$\pi(x)p(x, x+1) = 2^{-n}\frac{n!}{x!(n-x)!} \cdot \frac{n-x}{n}$$

$$= 2^{-n}\frac{n!}{(x+1)!(n-x-1)!} \cdot \frac{x+1}{n} = \pi(x+1)p(x+1, x)$$

However the following proof in words is simpler. Create X_0 by flipping coins, with heads = "in the left urn." The transition from X_0 to X_1 is done by picking a coin at random and then flipping it over. It should be clear that all 2^n outcomes of the coin tosses at time 1 are equally likely, so X_1 has the binomial distribution.

Example 1.28 (Three Machines, One Repairman). Suppose that an office has three machines that each break with probability .1 each day, but when there is at least one broken, then with probability 0.5 the repairman can fix one of them for use the next day. If we ignore the possibility of two machines breaking on the same day, then the number of working machines can be modeled as a birth and death chain with the following transition matrix:

	0	1	2	3
0	.5	.5	0	0
1	.05	.5	.45	0
2	0	.1	.5	.4
3	0	0	.3	.7

Rows 0 and 3 are easy to see. To explain row 1, we note that the state will only decrease by 1 if one machine breaks and the repairman fails to repair the one he is working on, an event of probability $(.1)(.5)$, while the state can only increase by 1 if he succeeds and there is no new failure, an event of probability $.5(.9)$. Similar reasoning shows $p(2, 1) = (.2)(.5)$ and $p(2, 3) = .5(.8)$.

To find the stationary distribution we use the recursive formula (1.12) to conclude that if $\pi(0) = c$ then

$$\pi(1) = \pi(0) \cdot \frac{p_0}{q_1} = c \cdot \frac{0.5}{0.05} = 10c$$

$$\pi(2) = \pi(1) \cdot \frac{p_1}{q_2} = 10c \cdot \frac{0.45}{0.1} = 45c$$

$$\pi(3) = \pi(2) \cdot \frac{p_2}{q_3} = 45c \cdot \frac{0.4}{0.3} = 60c$$

The sum of the π's is $116c$, so if we let $c = 1/116$ then we get

$$\pi(3) = \frac{60}{116}, \quad \pi(2) = \frac{45}{116}, \quad \pi(1) = \frac{10}{116}, \quad \pi(0) = \frac{1}{116}$$

There are many other Markov chains that are not birth and death chains but have stationary distributions that satisfy the detailed balance condition. A large number of possibilities are provided by

Example 1.29 (Random Walks on Graphs). A graph is described by giving two things: (i) a set of vertices V (which we suppose is a finite set) and (ii) an adjacency matrix $A(u, v)$, which is 1 if there is an edge connecting u and v and 0 otherwise. By convention we set $A(v, v) = 0$ for all $v \in V$.

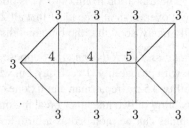

The degree of a vertex u is equal to the number of neighbors it has. In symbols,

$$d(u) = \sum_v A(u, v)$$

since each neighbor of u contributes 1 to the sum. To help explain the concept, we have indicated the degrees on our example. We write the degree this way to make it clear that

$$p(u, v) = \frac{A(u, v)}{d(u)} \qquad\qquad (*)$$

defines a transition probability. In words, if $X_n = u$, we jump to a randomly chosen neighbor of u at time $n + 1$.

It is immediate from $(*)$ that if c is a positive constant then $\pi(u) = cd(u)$ satisfies the detailed balance condition:

$$\pi(u)p(u, v) = cA(u, v) = cA(v, u) = \pi(v)p(u, v)$$

Thus, if we take $c = 1/\sum_u d(u)$, we have a stationary probability distribution. In the example $c = 1/40$.

For a concrete example, consider

Example 1.30 (Random Walk of a Knight on a Chess Board). A chess board is an 8 by 8 grid of squares. A knight moves by walking two steps in one direction and then one step in a perpendicular direction.

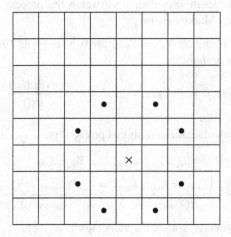

By patiently examining all of the possibilities, one sees that the degrees of the vertices are given by the following table. Lines have been drawn to make the symmetries more apparent.

2	3	4	4	4	4	3	2
3	4	6	6	6	6	4	3
4	6	8	8	8	8	6	4
4	6	8	8	8	8	6	4
4	6	8	8	8	8	6	4
4	6	8	8	8	8	6	4
3	4	6	6	6	6	4	3
2	3	4	4	4	4	3	2

The sum of the degrees is $4 \cdot 2 + 8 \cdot 3 + 20 \cdot 4 + 16 \cdot 6 + 16 \cdot 8 = 336$, so the stationary probabilities are the degrees divided by 336.

1.5.1 Reversibility

Let $p(i,j)$ be a transition probability with stationary distribution $\pi(i)$. Let X_n be a realization of the Markov chain starting from the stationary distribution, i.e., $P(X_0 = i) = \pi(i)$. The next result says that if we watch the process X_m, $0 \le m \le n$, backwards, then it is a Markov chain.

Theorem 1.15. *Fix n and let $Y_m = X_{n-m}$ for $0 \le m \le n$. Then Y_m is a Markov chain with transition probability*

$$\hat{p}(i,j) = P(Y_{m+1} = j|Y_m = i) = \frac{\pi(j)p(j,i)}{\pi(i)} \tag{1.13}$$

Proof. We need to calculate the conditional probability.

$$P(Y_{m+1} = i_{m+1}|Y_m = i_m, Y_{m-1} = i_{m-1} \ldots Y_0 = i_0)$$

$$= \frac{P(X_{n-(m+1)} = i_{m+1}, X_{n-m} = i_m, X_{n-m+1} = i_{m-1} \ldots X_n = i_0)}{P(X_{n-m} = i_m, X_{n-m+1} = i_{m-1} \ldots X_n = i_0)}$$

Using the Markov property, we see the numerator is equal to

$$\pi(i_{m+1})p(i_{m+1}, i_m)P(X_{n-m+1} = i_{m-1}, \ldots X_n = i_0|X_{n-m} = i_m)$$

Similarly the denominator can be written as

$$\pi(i_m)P(X_{n-m+1} = i_{m-1}, \ldots, X_n = i_0|X_{n-m} = i_m)$$

Dividing the last two formulas and noticing that the conditional probabilities cancel we have

$$P(Y_{m+1} = i_{m+1}|Y_m = i_m, \ldots Y_0 = i_0) = \frac{\pi(i_{m+1})p(i_{m+1}, i_m)}{\pi(i_m)}$$

This shows Y_m is a Markov chain with the indicated transition probability. □

The formula for the transition probability in (1.13), which is called the **dual transition probability**, may look a little strange, but it is easy to see that it works; i.e., the $\hat{p}(i,j) \geq 0$, and have

$$\sum_j \hat{p}(i,j) = \sum_j \pi(j)p(j,i)\pi(i) = \frac{\pi(i)}{\pi(i)} = 1$$

since $\pi p = \pi$. When π satisfies the detailed balance conditions:

$$\pi(i)p(i,j) = \pi(j)p(j,i)$$

the transition probability for the reversed chain,

$$\hat{p}(i,j) = \frac{\pi(j)p(j,i)}{\pi(i)} = p(i,j)$$

is the same as the original chain. In words, if we make a movie of the Markov chain $X_m, 0 \leq m \leq n$ starting from an initial distribution that satisfies the detailed balance condition and watch it backwards (i.e., consider $Y_m = X_{n-m}$ for $0 \leq m \leq n$), then we see a random process with the same distribution m. To help explain the concept,

1.5.2   The Metropolis–Hastings Algorithm

Our next topic is a method for generating samples from a distribution $\pi(x)$. It is named for two of the authors of the fundamental papers on the topic. One written by Nicholas Metropolis and two married couples with last names Rosenbluth and Teller (1953) and the other by Hastings (1970). This is a very useful tool for computing posterior distributions in Bayesian statistics (Tierney 1994), reconstructing images (Geman and Geman 1984), and investigating complicated models in statistical physics (Hammersley and Handscomb 1964). It would take us too far afield to describe these applications, so we will content ourselves to describe the simple idea that is the key to the method.

We begin with a Markov chain $q(x,y)$ that is the proposed jump distribution. A move is accepted with probability

$$r(x,y) = \min\left\{\frac{\pi(y)q(y,x)}{\pi(x)q(x,y)}, 1\right\}$$

so the transition probability

$$p(x,y) = q(x,y)r(x,y)$$

To check that π satisfies the detailed balance condition we can suppose that $\pi(y)q(y,x) > \pi(x)q(x,y)$. In this case

$$\pi(x)p(x,y) = \pi(x)q(x,y) \cdot 1$$

$$\pi(y)p(y,x) = \pi(y)q(y,x)\frac{\pi(x)q(x,y)}{\pi(y)q(y,x)} = \pi(x)q(x,y)$$

To generate one sample from $\pi(x)$ we run the chain for a long time so that it reaches equilibrium. To obtain many samples, we output the state at widely separated times. Of course there is an art of knowing how long is long enough to wait between outputting the state to have independent realizations. If we are interested in the expected value of a particular function, then (if the chain is irreducible and the state space is finite) Theorem 1.22 guarantees that

$$\frac{1}{n}\sum_{m=1}^{n}f(X_m) \to \sum_x f(x)\pi(x)$$

The Metropolis–Hastings algorithm is often used when space is continuous, but that requires a more sophisticated Markov chain theory, so we will use discrete examples to illustrate the method.

Example 1.31 (Geometric Distribution). Suppose $\pi(x) = \theta^x(1-\theta)$ for $x = 0,1,2,\ldots$. To generate the jumps we will use a symmetric random walk $q(x,x+1) = q(x,x-1) = 1/2$. Since q is symmetric $r(x,y) = \min\{1, \pi(y)/\pi(x)\}$. In this case if $x > 0$, $\pi(x-1) > \pi(x)$ and $\pi(x+1)/\pi(x) = \theta$ so

$$p(x,x-1) = 1/2 \quad p(x,x+1) = \theta/2 \quad p(x,x) = (1-\theta)/2.$$

When $x = 0$, $\pi(-1) = 0$ so

$$p(0,-1) = 0 \quad p(0,1) = \theta/2 \quad p(0,0) = 1 - (\theta/2).$$

To check reversibility we note that if $x \geq 0$ then

$$\pi(x)p(x,x+1) = \theta^x(1-\theta) \cdot \frac{\theta}{2} = \pi(x+1)p(x+1,x)$$

Here, as in most applications of the Metropolis–Hastings algorithm the choice of q is important. If θ is close to 1, then we would want to choose $q(x,x+i) = 1/2L+1$ for $-L \leq i \leq L$ where $L = O(1/(1-\theta))$ to make the chain move around the state space faster while not having too many steps rejected.

Example 1.32 (Binomial Distribution). Suppose $\pi(x)$ is Binomial(N,θ). In this case we can let $q(x,y) = 1/(N+1)$ for all $0 \leq x,y \leq N$. Since q is symmetric $r(x,y) = \min\{1,\pi(y)/\pi(x)\}$. This is closely related to the method of **rejection sampling**, in which one generates independent random variables U_i uniform on $\{0,1,\ldots,N\}$ and keep U_i with probability $\pi(U_i)/\pi^*$ where $\pi^* = \max_{0\leq x\leq n}\pi(x)$.

Example 1.33 (Two Dimensional Ising Model). The Metropolis–Hastings algorithm has its roots in statistical physics. A typical problem is the Ising model of ferromagnetism. Space is represented by a two dimensional grid $\Lambda = \{-L, \ldots L\}^2$. If we made the lattice three dimensional, we could think of the atoms in an iron bar. In reality each atom has a spin which can point in some direction, but we simplify by supposing that each spin can be up $+1$ or down -1. The state of the systems is a function $\xi : \Lambda \to \{-1, 1\}$ i.e., a point in the product space $\{-1, 1\}^\Lambda$. We say that points x and y in Λ are neighbors if y is one of the four points $x + (1, 0), x + (-1, 0), x + (0, 1), x + (0, -1)$. See the picture:

```
+ - + + + - -
- - - + + + -
+ - + + - - +
+ + - - y + -
+ - + y x y -
- - - + y + -
+ - + + - - +
```

Given an interaction parameter β, which is inversely proportional to the temperature, the equilibrium state is

$$\pi(x) = \frac{1}{Z(\beta)} \exp\left(\beta \sum_{x,y \sim x} \xi_x \xi_y \right)$$

where the sum is over all $x, y \in \Lambda$ with y a neighbor of x, and $Z(\beta)$ is a constant that makes the probabilities sum to one. At the boundaries of the square spins have only three neighbors. There are several options for dealing with this: (i) we consider the spins outside to be 0, or (ii) we could specify a fixed boundary condition such as all spins $+$.

The sum is largest in case (i) when all of the spins agree or in case (ii) when all spins are $+$. This configuration minimizes the energy $H = -\sum_{x,y \sim x} \eta_x \eta_y$ but there many more configurations one with a random mixture of $+$'s and $-$'s. It turns out that as β increases the system undergoes a phase transition from a random state with an almost equal number of $+$'s and $-$'s to one in which more than 1/2 of the spins point in the same direction.

$Z(\beta)$ is difficult to compute so it is fortunate that only the ratio of the probabilities appears in the Metropolis–Hastings recipe. For the proposed jump distribution we let $q(\xi, \xi^x) = 1/(2L + 1)^2$ if the two configurations ξ and ξ^x differ only at x. In this case the transition probability is

$$p(\xi, \xi^x) = q(\xi, \xi^x) \min\left\{ \frac{\pi(\xi^x)}{\pi(\xi)}, 1 \right\}$$

Note that the ratio $\pi(\xi^x)/\pi(\xi)$ is easy to compute because $Z(\beta)$ cancels out, as do all the terms in the sum that do not involve x and its neighbors. Since $\xi^x(x) = -\xi(x)$.

$$\frac{\pi(\xi^x)}{\pi(\xi)} = \exp\left(-2\beta \sum_{y \sim x} \xi_x \xi_y\right)$$

If x agrees with k of its four neighbors, the ratio is $\exp(-2(4-2k))$. In words $p(x, y)$ can be described by saying that we accept the proposed move with probability 1 if it lowers the energy and with probability $\pi(y)/\pi(x)$ if not.

Example 1.34 (Simulated Annealing). The Metropolis–Hastings algorithm can also be used to minimize complicated functions. Consider, for example, the traveling salesman problem, which is to find the shortest (or least expensive) route that allows one to visit all of the cities on a list. In this case the state space will be lists of cities, x and $\pi(x) = \exp(-\beta \ell(x))$ where $\ell(x)$ is the length of the tour. The proposal kernel q is chosen to modify the list in some way. For example, we might move a city to another place on the list or reverse the order of a sequence of cities. When β is large the stationary distribution will concentrate on optimal and near optimal tours. As in the Ising model, β is thought of as inverse temperature. The name derives from the fact that to force the chain to better solution we increase β (i.e., reduce the temperature) as we run the simulation. One must do this slowly or the process will get stuck in local minima. For more of simulated annealing, see Kirkpatrick et al. (1983)

1.5.3 Kolmogorow Cycle Condition

In this section we will state and prove a necessary and sufficient condition for an irreducible Markov chain to have a stationary distribution that satisfies the detailed balance condition. All irreducible two state chains have stationary distributions that satisfy detailed balance so we begin with the case of three states.

Example 1.35. Consider the chain with transition probability:

	1	2	3
1	$1-(a+d)$	a	d
2	e	$1-(b+e)$	b
3	c	f	$1-(c+f)$

and suppose that all entries in the matrix are positive. To satisfy detailed balance we must have

$$e\pi(2) = a\pi(1) \qquad f\pi(3) = b\pi(2) \qquad d\pi(1) = c\pi(3)$$

To construct a measure that satisfies detailed balance we take $\pi(1) = k$. From this it follows that

$$\pi(2) = k\frac{a}{e} \qquad \pi(3) = k\frac{ab}{ef} \qquad k = \pi(1) = k\frac{abc}{def}$$

From this we see that there is a stationary distribution satisfying (1.11) if and only if

$$abc = def \quad \text{or} \quad \frac{abc}{def}$$

To lead into the next result we note that this condition is

$$\frac{p(1,2) \cdot p(2,3) \cdot p(3,1)}{p(2,1) \cdot p(3,2) \cdot p(1,3)} = 1$$

i.e., the probabilities around the loop $1, 2, 3$ in either direction are equal.

Kolmogorov Cycle Condition Consider an irreducible Markov chain with state space S. We say that the cycle condition is satisfied if given a cycle of states $x_0, x_1, \ldots, x_n = x_0$ with $p(x_{i-1}, x_i) > 0$ for $1 \le i \le n$, we have

$$\prod_{i=1}^{n} p(x_{i-1}, x_i) = \prod_{i=1}^{n} p(x_i, x_{i-1}) \qquad (1.14)$$

Note that if this holds then $p(x_i, x_{i-1}) > 0$ for $1 \le i \le n$.

Theorem 1.16. *There is a stationary distribution that satisfies detailed balance if and only if (1.14) holds.*

Proof. We first show that if p has a stationary distribution that satisfies the detailed balance condition, then the cycle condition holds. Detailed balance implies

$$\frac{\pi(x_i)}{\pi(x_{i-1})} = \frac{p(x_{i-1}, x_i)}{p(x_i, x_{i-1})}.$$

Here and in what follows all of the transition probabilities we write down are positive. Taking the product from $i = 1, \ldots n$ and using the fact that $x_0 = x_n$ we have

$$1 = \prod_{i=1}^{n} \frac{\pi(x_i)}{\pi(x_{i-1})} = \prod_{i=1}^{n} \frac{p(x_{i-1}, x_i)}{q(x_i, x_{i-1})} \qquad (1.15)$$

To prove the converse, suppose that the cycle condition holds. Let $a \in S$ and set $\pi(a) = c$. For $b \ne a$ in S let $x_0 = a, x_1 \ldots x_k = b$ be a path from a to b with $p(x_{i-1}, x_i) > 0$ for $1 \le i \le k$ (and hence $p(x_i, x_{i-1}) > 0$ for $1 \le i \le k$. By the reasoning used to derive (1.15)

$$\pi(b) = c \prod_{j=1}^{k} \frac{q(x_{i-1}, x_i)}{q(x_i, x_{i-1})}$$

The first step is to show that $\pi(b)$ is well defined, i.e., is independent of the path chosen. Let $x_0 = a, \ldots x_k = b$ and $y_0 = a, \ldots y_\ell = b$ be two paths from a to b. Combine these to get a loop that begins and ends at a.

$$z_0 = x_0, \ldots z_k = x_k = y_\ell = b,$$

$$z_{k+1} = y_{\ell-1}, \ldots, z_{k+\ell} = y_0 = a$$

Since $z_{k+j} = y_{\ell-j}$ for $1 \le j \le \ell$, letting $h = \ell - j$ we have

$$1 = \prod_{h=1}^{k+\ell} \frac{p(z_{h-1}, z_h)}{p(z_h, z_{h-1})} = \prod_{i=1}^{k} \frac{p(x_{i-1}, x_i)}{p(x_i, x_{i-1})} \cdot \prod_{j=1}^{\ell} \frac{p(y_{\ell-j+1}, y_{\ell-j})}{p(y_{\ell-j}, y_{\ell-j+1})}$$

Changing variables $m = \ell - j + 1$ we have

$$\prod_{i=1}^{k} \frac{q(x_{i-1}, x_i)}{p(x_i, x_{i-1})} = \prod_{m=1}^{\ell} \frac{p(y_{m-1}, y_m)}{p(y_m, y_{m-1})}$$

This shows that the definition is independent of the path chosen. To check that π satisfies the detailed balance condition suppose $q(c, b) > 0$. Let $x_0 = a, \ldots, x_k = b$ be a path from a to b with $q(x_i, x_{i-1}) > 0$ for $1 \le i \le k$. If we let $x_{k+1} = c$, then since the definition is independent of path we have

$$\pi(b) = \prod_{i=1}^{k} \frac{p(x_{i-1}, x_i)}{p(x_i, x_{i-1})} \qquad \pi(c) = \prod_{i=1}^{k+1} \frac{q(x_{i-1}, x_i)}{p(x_i, x_{i-1})} = \pi(b) \frac{p(b, c)}{p(c, b)}$$

and the detailed balance condition is satisfied. $\qquad\qquad\qquad\qquad\qquad\square$

1.6 Limit Behavior

If y is a transient state, then by Lemma 1.11, $\sum_{n=1}^{\infty} p^n(x, y) < \infty$ for any initial state x and hence

$$p^n(x, y) \to 0$$

This means that we can restrict our attention to recurrent states and in view of the decomposition theorem, Theorem 1.8, to chains that consist of a single irreducible class of recurrent states. Our first example shows one problem that can prevent the convergence of $p^n(x, y)$.

Example 1.36 (Ehrenfest Chain (Continuation of 1.2)). For concreteness, suppose there are three balls. In this case the transition probability is

	0	**1**	**2**	**3**
0	0	3/3	0	0
1	1/3	0	2/3	0
2	0	2/3	0	1/3
3	0	0	3/3	0

In the second power of p the zero pattern is shifted:

	0	**1**	**2**	**3**
0	1/3	0	2/3	0
1	0	7/9	0	2/9
2	2/9	0	7/9	0
3	0	2/3	0	1/3

To see that the zeros will persist, note that if we have an odd number of balls in the left urn, then no matter whether we add or subtract one the result will be an even number. Likewise, if the number is even, then it will be odd on the next one step. This alternation between even and odd means that it is impossible to be back where we started after an odd number of steps. In symbols, if n is odd, then $p^n(x, x) = 0$ for all x.

To see that the problem in the last example can occur for multiples of any number N consider:

Example 1.37 (Renewal Chain). We will explain the name in Sect. 3.3. For the moment we will use it to illustrate "pathologies." Let f_k be a distribution on the positive integers and let $p(0, k - 1) = f_k$. For states $i > 0$ we let $p(i, i - 1) = 1$. In words the chain jumps from 0 to $k - 1$ with probability f_k and then walks back to 0 one step at a time. If $X_0 = 0$ and the jump is to $k - 1$, then it returns to 0 at time k. If say $f_5 = f_{15} = 1/2$, then $p^n(0, 0) = 0$ unless n is a multiple of 5.

The **period** of a state is the largest number that will divide all the $n \geq 1$ for which $p^n(x, x) > 0$. That is, it is the greatest common divisor of $I_x = \{n \geq 1 : p^n(x, x) > 0\}$. To check that this definition works correctly, we note that in Example 1.36, $\{n \geq 1 : p^n(x, x) > 0\} = \{2, 4, \ldots\}$, so the greatest common divisor is 2. Similarly, in Example 1.37, $\{n \geq 1 : p^n(x, x) > 0\} = \{5, 10, \ldots\}$, so the greatest common divisor is 5. As the next example shows, things aren't always so simple.

Example 4.4 (Triangle and Square). Consider the transition matrix:

	-2	-1	0	1	2	3
-2	0	0	1	0	0	0
-1	1	0	0	0	0	0
0	0	0	0.5	0	0.5	0
1	0	0	0	0	0	1
2	0	0	0	0	0	1
3	0	0	1	0	0	0

In words, from 0 we are equally likely to go to 1 or -1. From -1 we go with probability one to -2 and then back to 0, from 1 we go to 2 then to 3 and back to 0. The name refers to the fact that $0 \to -1 \to -2 \to 0$ is a triangle and $0 \to 1 \to 2 \to 3 \to 0$ is a square.

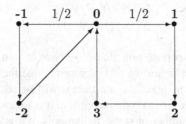

Clearly, $p^3(0,0) > 0$ and $p^4(0,0) > 0$ so $3, 4 \in I_0$ and hence $d_0 = 1$. To determine the periods of the other states it is useful to know.

Lemma 1.17. *If $\rho_{xy} > 0$ and $\rho_{yx} > 0$, then x and y have the same period.*

Why is this true? The short answer is that if the two states have different periods, then by going from x to y, from y to y in the various possible ways, and then from y to x, we will get a contradiction.

Proof. Suppose that the period of x is c, while the period of y is $d < c$. Let k be such that $p^k(x, y) > 0$ and let m be such that $p^m(y, x) > 0$. Since

$$p^{k+m}(x,x) \geq p^k(x,y)p^m(y,x) > 0$$

we have $k + m \in I_x$. Since x has period c, $k + m$ must be a multiple of c. Now let ℓ be any integer with $p^\ell(y,y) > 0$. Since

$$p^{k+\ell+m}(x,x) \geq p^k(x,y)p^\ell(y,y)p^m(y,x) > 0$$

$k + \ell + m \in I_x$, and $k + \ell + m$ must be a multiple of c. Since $k + m$ is itself a multiple of c, this means that ℓ is a multiple of c. Since $\ell \in I_y$ was arbitrary, we have shown that c is a divisor of every element of I_y, but $d < c$ is the greatest common divisor, so we have a contradiction. $\square$

With Lemma 1.17 in hand, we can show aperiodicity in many examples by using.

Lemma 1.18. *If $p(x,x) > 0$, then x has period 1.*

Proof. If $p(x,x) > 0$, then $1 \in I_x$, so the greatest common divisor is 1. □

This result by itself is enough to show that all states in the weather chain (Example 1.3), social mobility (Example 1.4), and brand preference chain (Example 1.5) are aperiodic. Combining it with Lemma 1.17 easily settles the question for the inventory chain (Example 1.6)

$$
\begin{array}{c|cccccc}
 & 0 & 1 & 2 & 3 & 4 & 5 \\
\hline
0 & 0 & 0 & .1 & .2 & .4 & .3 \\
1 & 0 & 0 & .1 & .2 & .4 & .3 \\
2 & .3 & .4 & .3 & 0 & 0 & 0 \\
3 & .1 & .2 & .4 & .3 & 0 & 0 \\
4 & 0 & .1 & .2 & .4 & .3 & 0 \\
5 & 0 & 0 & .1 & .2 & .4 & .3 \\
\end{array}
$$

Since $p(x,x) > 0$ for $x = 2,3,4,5$, Lemma 1.18 implies that these states are aperiodic. Since this chain is irreducible it follows from Lemma 1.17 that 0 and 1 are aperiodic.

Consider now the basketball chain (Example 1.10):

$$
\begin{array}{c|cccc}
 & \textbf{HH} & \textbf{HM} & \textbf{MH} & \textbf{MM} \\
\hline
\textbf{HH} & 3/4 & 1/4 & 0 & 0 \\
\textbf{HM} & 0 & 0 & 2/3 & 1/3 \\
\textbf{MH} & 2/3 & 1/3 & 0 & 0 \\
\textbf{MM} & 0 & 0 & 1/2 & 1/2 \\
\end{array}
$$

Lemma 1.18 implies that **HH** and **MM** are aperiodic. Since this chain is irreducible it follows from Lemma 1.17 that **HM** and **MH** are aperiodic.

We now come to the main results of the chapter. We first list the assumptions. All of these results hold when S is finite or infinite.

- $I : p$ is irreducible
- $A :$ aperiodic, all states have period 1
- $R :$ all states are recurrent
- $S :$ there is a stationary distribution π

Theorem 1.19 (Convergence Theorem). *Suppose I, A, S. Then as $n \to \infty$, $p^n(x,y) \to \pi(y)$.*

The next result describes the "limiting fraction of time we spend in each state."

Theorem 1.20 (Asymptotic Frequency). *Suppose I and R. If $N_n(y)$ be the number of visits to y up to time n, then*

$$
\frac{N_n(y)}{n} \to \frac{1}{E_y T_y}
$$

We will see later that when the state space is infinite we may have $E_y T_y = \infty$ in which case the limit is 0. As a corollary we get the following.

Theorem 1.21. *If I and S hold, then*

$$\pi(y) = 1/E_y T_y$$

and hence the stationary distribution is unique.

In the next two examples we will be interested in the long run cost associated with a Markov chain. For this, we will need the following extension of Theorem 1.20.

Theorem 1.22. *Suppose I, S, and $\sum_x |f(x)|\pi(x) < \infty$ then*

$$\frac{1}{n}\sum_{m=1}^{n} f(X_m) \to \sum_x f(x)\pi(x)$$

Note that only Theorem 1.19 requires aperiodicity. Taking $f(x) = 1$ if $x = y$ and 0 otherwise in Theorem 1.22 gives Theorem 1.20. If we then take expected value we have

Theorem 1.23. *Suppose I, S.*

$$\frac{1}{n}\sum_{m=1}^{n} p^m(x, y) \to \pi(y)$$

Thus while the sequence $p^m(x, y)$ will not converge in the periodic case, the average of the first n values will.

To illustrate the use of these results, we consider

Example 1.38 (Repair Chain (Continuation of 1.7)). A machine has three critical parts that are subject to failure, but can function as long as two of these parts are working. When two are broken, they are replaced and the machine is back to working order the next day. Declaring the state space to be the parts that are broken $\{0, 1, 2, 3, 12, 13, 23\}$, we arrived at the following transition matrix:

	0	1	2	3	12	13	23
0	.93	.01	.02	.04	0	0	0
1	0	.94	0	0	.02	.04	0
2	0	0	.95	0	.01	0	.04
3	0	0	0	.97	0	.01	.02
12	1	0	0	0	0	0	0
13	1	0	0	0	0	0	0
23	1	0	0	0	0	0	0

and we asked: If we are going to operate the machine for 1800 days (about 5 years), then how many parts of types 1, 2, and 3 will we use?

To find the stationary distribution we look at the last row of

$$
\begin{pmatrix}
-.07 & .01 & .02 & .04 & 0 & 0 & 1 \\
0 & -.06 & 0 & 0 & .02 & .04 & 1 \\
0 & 0 & -.05 & 0 & .01 & 0 & 1 \\
0 & 0 & 0 & -.03 & 0 & .01 & 1 \\
1 & 0 & 0 & 0 & -1 & 0 & 1 \\
1 & 0 & 0 & 0 & 0 & -1 & 1 \\
1 & 0 & 0 & 0 & 0 & 0 & 1
\end{pmatrix}^{-1}
$$

where after converting the results to fractions we have

$$\pi(0) = 3000/8910$$

$$\pi(1) = 500/8910 \quad \pi(2) = 1200/8910 \quad \pi(3) = 4000/8910$$

$$\pi(12) = 22/8910 \quad \pi(13) = 60/8910 \quad \pi(23) = 128/8910$$

We use up one part of type 1 on each visit to 12 or to 13, so on the average we use 82/8910 of a part per day. Over 1800 days we will use an average of $1800 \cdot 82/8910 = 16.56$ parts of type 1. Similarly type 2 and type 3 parts are used at the long run rates of 150/8910 and 188/8910 per day, so over 1800 days we will use an average of 30.30 parts of type 2 and 37.98 parts of type 3.

Example 1.39 (Inventory Chain (Continuation of 1.6)). We have an electronics store that sells a videogame system, with the potential for sales of 0, 1, 2, or 3 of these units each day with probabilities .3, .4, .2, and .1. Each night at the close of business new units can be ordered which will be available when the store opens in the morning.

As explained earlier if X_n is the number of units on hand at the end of the day then we have the following transition matrix:

	0	1	2	3	4	5
0	0	0	.1	.2	.4	.3
1	0	0	.1	.2	.4	.3
2	.3	.4	.3	0	0	0
3	.1	.2	.4	.3	0	0
4	0	.1	.2	.4	.3	0
5	0	0	.1	.2	.4	.3

In the first section we asked the question:

Q. Suppose we make $12 profit on each unit sold but it costs $2 a day to store items. What is the long-run profit per day of this inventory policy?

The first thing we have to do is to compute the stationary distribution. The last row of

$$\begin{pmatrix} -1 & 0 & .1 & .2 & .4 & 1 \\ 0 & -1 & .1 & .2 & .4 & 1 \\ .3 & .4 & -.7 & 0 & 0 & 1 \\ .1 & .2 & .4 & -.7 & 0 & 1 \\ 0 & .1 & .2 & .4 & -.7 & 1 \\ 0 & 0 & .1 & .2 & .4 & .1 \end{pmatrix}^{-1}$$

has entries

$$\frac{1}{9740}(885, 1516, 2250, 2100, 1960, 1029)$$

We first compute the average number of units sold per day. To do this we note that the average demand per day is

$$ED = 0.4 \cdot 1 + 0.2 \cdot 2 + 0.1 \cdot 3 = 1.1$$

so we only have to compute the average number of lost sales per day. This only occurs when $X_n = 2$ and $D_{n+1} = 3$ and in this case we lose one sale so the lost sales per day is

$$\pi(2)P(D_{n+1} = 3) = \frac{2250}{9740} \cdot 0.1 = 0.0231$$

Thus the average number of sales per day is $1.1 - 0.023 = 1.077$ for a profit of $1.077 \cdot 12 = 12.94$.

Taking $f(k) = 2k$ in Theorem 1.22 we see that in the long our average holding costs per day are

$$\frac{1}{9740}(1516 \cdot 2 + 2250 \cdot 4 + 2100 \cdot 6 + 1960 \cdot 8 + 1029 \cdot 10) = 5.20.$$

1.7 Returns to a Fixed State

In this section we will prove Theorems 1.20, 1.21, and 1.22. The key is to look at the times the chain returns to a fixed state.

Theorem 1.20. *Suppose p is irreducible and recurrent. Let $N_n(y)$ be the number of visits to y at times $\leq n$. As $n \to \infty$*

$$\frac{N_n(y)}{n} \to \frac{1}{E_y T_y}$$

Why is this true? Suppose first that we start at y. The times between returns, $t_1, t_2, \ldots$ are independent and identically distributed so the strong law of large numbers for nonnegative random variables implies that the time of the kth return to y, $R(k) = \min\{n \geq 1 : N_n(y) = k\}$, has

$$\frac{R(k)}{k} \to E_y T_y \leq \infty \tag{1.16}$$

If we do not start at y, then $t_1 < \infty$ and $t_2, t_3, \ldots$ are independent and identically distributed and we again have (1.16). Writing $a_k \sim b_k$ when $a_k/b_k \to 1$ we have $R(k) \sim k E_y T_y$. Taking $k = n/E_y T_y$ we see that there are about $n/E_y T_y$ returns by time n.

Proof. We have already shown (1.16). To turn this into the desired result, we note that from the definition of $R(k)$ it follows that $R(N_n(y)) \leq n < R(N_n(y) + 1)$. Dividing everything by $N_n(y)$ and then multiplying and dividing on the end by $N_n(y) + 1$, we have

$$\frac{R(N_n(y))}{N_n(y)} \leq \frac{n}{N_n(y)} < \frac{R(N_n(y) + 1)}{N_n(y) + 1} \cdot \frac{N_n(y) + 1}{N_n(y)}$$

Letting $n \to \infty$, we have $n/N_n(y)$ trapped between two things that converge to $E_y T_y$, so

$$\frac{n}{N_n(y)} \to E_y T_y$$

and we have proved the desired result. $\qquad\qquad\qquad\qquad\qquad\qquad\qquad\quad\square$

From the last result we immediately get

Theorem 1.21. *If p is an irreducible and has stationary distribution π, then*

$$\pi(y) = 1/E_y T_y$$

Proof. Suppose X_0 has distribution π. From Theorem 1.20 it follows that

$$\frac{N_n(y)}{n} \to \frac{1}{E_y T_y}$$

Taking expected value and using the fact that $N_n(y) \leq n$, it can be shown that this implies

$$\frac{E_\pi N_n(y)}{n} \to \frac{1}{E_y T_y}$$

but since π is a stationary distribution $E_\pi N_n(y) = n\pi(y)$. $\qquad\qquad\qquad\quad\square$

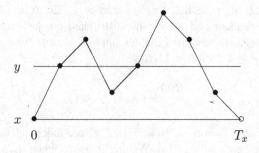

Fig. 1.2 Picture of the cycle trick

A corollary of this result is that for an irreducible chain if there is a stationary distribution it is unique. Our next topic is the existence of stationary measures

Theorem 1.24. *Suppose p is irreducible and recurrent. Let $x \in S$ and let $T_x = \inf\{n \geq 1 : X_n = x\}$.*

$$\mu_x(y) = \sum_{n=0}^{\infty} P_x(X_n = y, T_x > n)$$

defines a stationary measure with $0 < \mu_x(y) < \infty$ for all y.

Why is this true? This is called the "cycle trick." $\mu_x(y)$ is the expected number of visits to y in $\{0, \dots, T_x - 1\}$. Multiplying by p moves us forward one unit in time so $\mu_x p(y)$ is the expected number of visits to y in $\{1, \dots, T_x\}$. Since $X(T_x) = X_0 = x$ it follows that $\mu_x = \mu_x p$ (Fig. 1.2).

Proof. To formalize this intuition, let $\bar{p}_n(x, y) = P_x(X_n = y, T_x > n)$ and interchange sums to get

$$\sum_{y} \mu_x(y) p(y, z) = \sum_{n=0}^{\infty} \sum_{y} \bar{p}_n(x, y) p(y, z)$$

Case 1. Consider the generic case first: $z \neq x$.

$$\sum_{y} \bar{p}_n(x, y) p(y, z) = \sum_{y} P_x(X_n = y, T_x > n, X_{n+1} = z)$$

$$= P_x(T_x > n + 1, X_{n+1} = z) = \bar{p}_{n+1}(x, z)$$

Here the second equality holds since the chain must be somewhere at time n, and the third is just the definition of $\bar{p}_{n+1}$. Summing from $n = 0$ to ∞, we have

$$\sum_{n=0}^{\infty} \sum_{y} \bar{p}_n(x, y) p(y, z) = \sum_{n=0}^{\infty} \bar{p}_{n+1}(x, z) = \mu_x(z)$$

since $\bar{p}_0(x, z) = 0$.

Case 2. Now suppose that $z = x$. Reasoning as above we have

$$\sum_y \bar{p}_n(x, y) p(y, x) = \sum_y P_x(X_n = y, T_x > n, X_{n+1} = x) = P_x(T_x = n + 1)$$

Summing from $n = 0$ to ∞ we have

$$\sum_{n=0}^{\infty} \sum_y \bar{p}_n(x, y) p(y, x) = \sum_{n=0}^{\infty} P_x(T_x = n + 1) = 1 = \mu_x(x)$$

since $P_x(T_x = 0) = 0$.

To check $\mu_x(y) < \infty$, we note that $\mu_x(x) = 1$ and

$$1 = \mu_x(x) = \sum_z \mu_x(z) p^n(z, x) \geq \mu_x(y) p^n(y, x)$$

so if we pick n with $p^n(y, x) > 0$ then we conclude $\mu_x(y) < \infty$.

To prove that $\mu_x(y) > 0$ we note that this is trivial for $y = x$ the point used to define the measure. For $y \neq x$, we borrow an idea from Theorem 1.5. Let $K = \min\{k : p^k(x, y) > 0\}$. Since $p^K(x, y) > 0$ there must be a sequence $y_1, \ldots y_{K-1}$ so that

$$p(x, y_1) p(y_1, y_2) \cdots p(y_{K-1}, y) > 0$$

Since K is minimal all the $y_i \neq y$, so $P_x(X_K = y, T_x > K) > 0$ and hence $\mu_x(y) > 0$.

$\square$

Our next step is to prove

Theorem 1.22. *Suppose p is irreducible, has stationary distribution π, and $\sum_x |f(x)| \pi(x) < \infty$ then*

$$\frac{1}{n} \sum_{m=1}^{n} f(X_m) \to \sum_x f(x) \pi(x)$$

The key idea here is that by breaking the path at the return times to x we get a sequence of random variables to which we can apply the law of large numbers.

Sketch of Proof. Suppose that the chain starts at x. Let $T_0 = 0$ and $T_k = \min\{n > T_{k-1} : X_n = x\}$ be the time of the kth return to x. By the strong Markov property, the random variables

$$Y_k = \sum_{m=T_{k-1}+1}^{T_k} f(X_m)$$

are independent and identically distributed. By the cycle trick in the proof of Theorem 1.24

$$EY_k = \sum_x \mu_x(y) f(y)$$

Using the law of large numbers for i.i.d. variables

$$\frac{1}{L} \sum_{m=1}^{T_L} f(X_m) = \frac{1}{L} \sum_{k=1}^{L} Y_k \to \sum_x \mu_x(y) f(y)$$

Taking $L = N_n(x) = \max\{k : T_k \le n\}$ and ignoring the contribution from the last incomplete cycle $(N_n(x), n]$

$$\frac{1}{n} \sum_{m=1}^{n} f(X_m) \approx \frac{N_n(x)}{n} \cdot \frac{1}{N_n(x)} \sum_{k=1}^{N_n(x)} Y_k$$

Using Theorem 1.20 and the law of large numbers the above

$$\to \frac{1}{E_x T_x} \sum_y \mu_x(y) f(y) = \sum_y \pi(y) f(y)$$

$\square$

1.8   Proof of the Convergence Theorem*

To prepare for the proof of the convergence theorem, Theorem 1.19, we need the following:

Lemma 1.25. *If there is a stationary distribution, then all states y that have $\pi(y) > 0$ are recurrent.*

Proof. Lemma 1.12 tells us that $E_x N(y) = \sum_{n=1}^{\infty} p^n(x, y)$, so

$$\sum_x \pi(x) E_x N(y) = \sum_x \pi(x) \sum_{n=1}^{\infty} p^n(x, y)$$

Interchanging the order of summation and then using $\pi p^n = \pi$, the above

$$= \sum_{n=1}^{\infty} \sum_x \pi(x) p^n(x, y) = \sum_{n=1}^{\infty} \pi(y) = \infty$$

since $\pi(y) > 0$. Using Lemma 1.11 now gives $E_x N(y) = \rho_{xy}/(1 - \rho_{yy})$, so

$$\infty = \sum_x \pi(x) \frac{\rho_{xy}}{1 - \rho_{yy}} \le \frac{1}{1 - \rho_{yy}}$$

the second inequality following from the facts that $\rho_{xy} \le 1$ and π is a probability measure. This shows that $\rho_{yy} = 1$, i.e., y is recurrent.   □

The next ingredient is

Lemma 1.26. *If x has period 1, i.e., the greatest common divisor I_x is 1, then there is a number n_0 so that if $n \ge n_0$, then $n \in I_x$. In words, I_x contains all of the integers after some value n_0.*

To prove this we begin by proving

Lemma 1.27. *I_x is closed under addition. That is, if $i, j \in I_x$, then $i + j \in I_x$.*

Proof. If $i, j \in I_x$, then $p^i(x, x) > 0$ and $p^j(x, x) > 0$ so

$$p^{i+j}(x, x) \ge p^i(x, x) p^j(x, x) > 0$$

and hence $i + j \in I_x$.   □

Using this we see that in the triangle and square example

$$I_0 = \{3, 4, 6, 7, 8, 9, 10, 11, \ldots\}$$

Note that in this example once we have three consecutive numbers (e.g., 6, 7, 8) in I_0 then $6 + 3, 7 + 3, 8 + 3 \in I_0$ and hence I_0 will contain all the integers $n \ge 6$.

For another unusual example consider the renewal chain (Example 1.37) with $f_5 = f_{12} = 1/2$. $5, 12 \in I_0$ so using Lemma 1.27

$$I_0 = \{5, 10, 12, 15, 17, 20, 22, 24, 25, 27, 29, 30, 32,$$
$$34, 35, 36, 37, 39, 40, 41, 42, 43, \ldots\}$$

To check this note that 5 gives rise to $10 = 5 + 5$ and $17 = 5 + 12$, 10 to 15 and 22, 12 to 17 and 24, etc. Once we have five consecutive numbers in I_0, here 39–43, we have all the rest. The last two examples motivate the following.

Proof. We begin by observing that it is enough to show that I_x will contain two consecutive integers: k and $k + 1$. For then it will contain $2k, 2k + 1, 2k + 2$, and $3k, 3k + 1, 3k + 2, 3k + 3$, or in general $jk, jk + 1, \ldots jk + j$. For $j \ge k - 1$ these blocks overlap and no integers are left out. In the last example $24, 25 \in I_0$ implies $48, 49, 50 \in I_0$ which implies $72, 73, 74, 75 \in I_0$ and $96, 97, 98, 99, 100 \in I_0$, so we know the result holds for $n_0 = 96$. In fact it actually holds for $n_0 = 34$ but it is not important to get a precise bound.

To show that there are two consecutive integers, we cheat and use a fact from number theory: if the greatest common divisor of a set I_x is 1, then there are integers $i_1, \ldots i_m \in I_x$ and (positive or negative) integer coefficients c_i so that $c_1 i_1 + \cdots + c_m i_m = 1$. Let $a_i = c_i^+$ and $b_i = (-c_i)^+$. In words the a_i are the positive coefficients and the b_i are -1 times the negative coefficients. Rearranging the last equation gives

$$a_1 i_1 + \cdots + a_m i_m = (b_1 i_1 + \cdots + b_m i_m) + 1$$

and using Lemma 1.27 we have found our two consecutive integers in I_x.   □

With Lemmas 1.25 and 1.26 in hand we are ready to tackle the proof of:

Theorem 1.23 (Convergence Theorem). *Suppose p is irreducible, aperiodic, and has stationary distribution π. Then as $n \to \infty$, $p^n(x, y) \to \pi(y)$.*

Proof. Let S be the state space for p. Define a transition probability $\bar{p}$ on $S \times S$ by

$$\bar{p}((x_1, y_1), (x_2, y_2)) = p(x_1, x_2)p(y_1, y_2)$$

In words, each coordinate moves independently.

Step 1. We will first show that if p is aperiodic and irreducible then $\bar{p}$ is irreducible. Since p is irreducible, there are K, L, so that $p^K(x_1, x_2) > 0$ and $p^L(y_1, y_2) > 0$. Since x_2 and y_2 have period 1, it follows from Lemma 1.26 that if M is large, then $p^{L+M}(x_2, x_2) > 0$ and $p^{K+M}(y_2, y_2) > 0$, so

$$\bar{p}^{K+L+M}((x_1, y_1), (x_2, y_2)) > 0$$

Step 2. Since the two coordinates are independent $\bar{\pi}(a, b) = \pi(a)\pi(b)$ defines a stationary distribution for $\bar{p}$, and Lemma 1.25 implies that all states are recurrent for $\bar{p}$. Let (X_n, Y_n) denote the chain on $S \times S$, and let T be the first time that the two coordinates are equal, i.e., $T = \min\{n \geq 0 : X_n = Y_n\}$. Let $V_{(x,x)} = \min\{n \geq 0 : X_n = Y_n = x\}$ be the time of the first visit to (x, x). Since $\bar{p}$ is irreducible and recurrent, $V_{(x,x)} < \infty$ with probability one. Since $T \leq V_{(x,x)}$ for any x we must have

$$P(T < \infty) = 1. \tag{1.17}$$

Step 3. By considering the time and place of the first intersection and then using the Markov property we have

$$P(X_n = y, T \leq n) = \sum_{m=1}^{n} \sum_x P(T = m, X_m = x, X_n = y)$$

$$= \sum_{m=1}^{n} \sum_x P(T = m, X_m = x)P(X_n = y | X_m = x)$$

$$= \sum_{m=1}^{n} \sum_{x} P(T = m, Y_m = x) P(Y_n = y | Y_m = x)$$

$$= P(Y_n = y, T \le n)$$

Step 4. To finish up we observe that since the distributions of X_n and Y_n agree on $\{T \le n\}$

$$|P(X_n = y) - P(Y_n = y)| \le P(X_n = y, T > n) + P(Y_n = y, T > n)$$

and summing over y gives

$$\sum_{y} |P(X_n = y) - P(Y_n = y)| \le 2P(T > n)$$

If we let $X_0 = x$ and let Y_0 have the stationary distribution π, then Y_n has distribution π, and using (1.17) it follows that

$$\sum_{y} |p^n(x, y) - \pi(y)| \le 2P(T > n) \to 0$$

proving the convergence theorem.   □

1.9   Exit Distributions

To motivate developments, we begin with an example.

Example 1.40 (Two Year College). At a local two year college, 60 % of freshmen become sophomores, 25 % remain freshmen, and 15 % drop out. 70 % of sophomores graduate and transfer to a four year college, 20 % remain sophomores and 10 % drop out. What fraction of new students eventually graduate?

We use a Markov chain with state space 1 = freshman, 2 = sophomore, G = graduate, D = dropout. The transition probability is

	1	2	G	D
1	0.25	0.6	0	0.15
2	0	0.2	0.7	0.1
G	0	0	1	0
D	0	0	0	1

Let $h(x)$ be the probability that a student currently in state x eventually graduates. By considering what happens on one step

$$h(1) = 0.25h(1) + 0.6h(2)$$

$$h(2) = 0.2h(2) + 0.7$$

To solve we note that the second equation implies $h(2) = 7/8$ and then the first that

$$h(1) = \frac{0.6}{0.75} \cdot \frac{7}{8} = 0.7$$

Example 1.41 (Tennis). In tennis the winner of a game is the first player to win four points, unless the score is $4 - 3$, in which case the game must continue until one player is ahead by two points and wins the game. Suppose that the server win the point with probability 0.6 and successive points are independent. What is the probability the server will win the game if the score is tied 3–3? if she is ahead by one point? Behind by one point?

We formulate the game as a Markov chain in which the state is the difference of the scores. The state space is $2, 1, 0, -1, -2$ with 2 (win for server) and -2 (win for opponent). The transition probability is

	2	1	0	-1	-2
2	1	0	0	0	0
1	.6	0	.4	0	0
0	0	.6	0	.4	0
-1	0	0	.6	0	.4
-2	0	0	0	0	1

If we let $h(x)$ be the probability of the server winning when the score is x, then

$$h(x) = \sum_y p(x, y)h(y)$$

with $h(2) = 1$ and $h(-2) = 0$. This gives us three equations in three unknowns

$$h(1) = .6 + .4h(0)$$

$$h(0) = .6h(1) + .4h(-1)$$

$$h(-1) = .6h(0)$$

Using the first and third equations in the second we have

$$h(0) = .6(.6 + .4h(0)) + .4(.6h(0)) = .36 + .48h(0)$$

so we have $h(0) = 0.36/0.52 = 0.6923$.

The last computation uses special properties of this example. To introduce a general approach, we rearrange the equations to get

$$h(1) - .4h(0) + 0h(-1) = .6$$

$$-.6h(1) + h(0) - .4h(-1) = 0$$

$$0h(1) - .6h(0) + h(-1) = 0$$

which can be written in matrix form as

$$\begin{pmatrix} 1 & -.4 & 0 \\ -.6 & 1 & -.4 \\ 0 & -.6 & 1 \end{pmatrix} \begin{pmatrix} h(1) \\ h(0) \\ h(-1) \end{pmatrix} = \begin{pmatrix} .6 \\ 0 \\ 0 \end{pmatrix}$$

Let $C = \{1, 0, -1\}$ be the nonabsorbing states and let $r(x, y)$ the restriction of p to $x, y \in C$ (i.e., the 3×3 matrix inside the black lines in the transition probability). In this notation then the matrix above is $I - r$, while the right-hand side is $v_i = p(i, a)$. Solving gives

$$\begin{pmatrix} h(1) \\ h(0) \\ h(-1) \end{pmatrix} = (I - r)^{-1} \begin{pmatrix} .6 \\ 0 \\ 0 \end{pmatrix} = \begin{pmatrix} .8769 \\ .6923 \\ .4154 \end{pmatrix}$$

General Solution Suppose that the server wins each point with probability w. If the game is tied then after two points, the server will have won with probability w^2, lost with probability $(1 - w)^2$, and returned to a tied game with probability $2w(1 - w)$, so $h(0) = w^2 + 2w(1 - w)h(0)$. Since $1 - 2w(1 - w) = w^2 + (1 - w)^2$, solving gives

$$h(0) = \frac{w^2}{w^2 + (1 - w)^2}$$

Figure 1.3 graphs this function.

Having worked two examples, it is time to show that we have computed the right answer. In some cases we will want to guess and verify the answer. In those situations it is nice to know that the solution is unique. The next result proves this. Given a set F let $V_F = \min\{n \geq 0 : X_n \in F\}$.

Theorem 1.28. *Consider a Markov chain with state space S. Let A and B be subsets of S, so that $C = S - (A \cup B)$ is finite. Suppose $h(a) = 1$ for $a \in A$, $h(b) = 0$ for $b \in B$, and that for $x \in C$ we have*

$$h(x) = \sum_y p(x, y)h(y) \tag{1.18}$$

If $P_x(V_A \wedge V_B < \infty) > 0$ for all $x \in C$, then $h(x) = P_x(V_a < V_b)$.

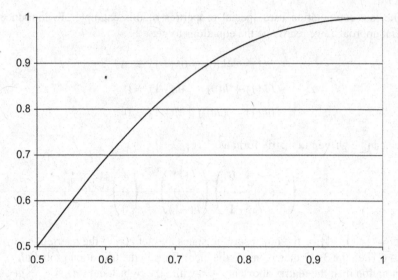

Fig. 1.3 Probability of winning a game in tennis as a function of the probability of winning a point

Proof. Let $T = V_A \wedge V_B$. It follows from Lemma 1.3 that $P_x(T < \infty) = 1$ for all $x \in C$. (1.18) implies that $h(x) = E_x h(X_1)$ when $x \in C$. The Markov property implies

$$h(x) = E_x h(X_{T \wedge n}).$$

We have to stop at time T because the equation is not assumed to be valid for $x \in A \cup B$. Since h is bounded with $h(a) = 1$ for $a \in A$, and $h(b) = 0$ for $b \in B$, it follows from the bounded convergence theorem A.10 that $E_x h(X_{T \wedge n}) \to P_x(V_A < V_B)$ which gives the desired result.   □

Example 1.42 (Matching Pennies). Bob, who has 15 pennies, and Charlie, who has 10 pennies, decide to play a game. They each flip a coin. If the two coins match, Bob gets the two pennies (for a profit of 1). If the two coins are different, then Charlie gets the two pennies. They quit when someone has all of the pennies. What is the probability Bob will win the game?

The answer will turn out to be 15/25, Bob's fraction of the total supply of pennies. To explain this, let X_n be the number of pennies Bob has after n plays. X_n is a fair game, i.e., $x = E_x X_1$, or in words the expected number of pennies Bob has is constant in time. Let

$$V_y = \min\{n \geq 0 : X_n = y\}$$

be the time of the first visit to y. Taking a leap of faith the expected number he has at the end of the game should be the same as at the beginning so

$$x = N P_x(V_N < V_0) + 0 P_x(V_0 < V_n)$$

and solving gives

$$P_x(V_N < V_0) = x/N \quad \text{for } 0 \le x \le N \tag{1.19}$$

To prove this note that by considering what happens on the first step

$$h(x) = \frac{1}{2}h(x+1) + \frac{1}{2}h(x-1)$$

Multiplying by 2 and rearranging

$$h(x+1) - h(x) = h(x) - h(x-1)$$

or in words, h has constant slope. Since $h(0) = 0$ and $h(N) = 1$ the slope must be $1/N$ and we must have $h(x) = x/N$.

The reasoning in the last example can be used to study Example 1.9.

Example 1.43 (Wright–Fisher Model with No Mutation). The state space is $S = \{0, 1, \ldots N\}$ and the transition probability is

$$p(x, y) = \binom{N}{y} \left(\frac{x}{N}\right)^y \left(\frac{N-x}{N}\right)^{N-y}$$

The right-hand side is the binomial($N, x/N$) distribution, i.e., the number of successes in N trials when success has probability x/N, so the mean number of successes is x. From this it follows that if we define $h(x) = x/N$, then

$$h(x) = \sum_y p(x, y)h(y)$$

Taking $a = N$ and $b = 0$, we have $h(a) = 1$ and $h(b) = 0$. Since $P_x(V_a \wedge V_b < \infty) > 0$ for all $0 < x < N$, it follows from Lemma 1.28 that

$$P_x(V_N < V_0) = x/N \tag{1.20}$$

i.e., the probability of fixation to all A's is equal to the fraction of the genes that are A.

Our next topic is non-fair games.

Example 1.44 (Gambler's Ruin). Consider a gambling game in which on any turn you win \$1 with probability $p \ne 1/2$ or lose \$1 with probability $1 - p$. Suppose further that you will quit playing if your fortune reaches \$$N$. Of course, if your fortune reaches \$0, then the casino makes you stop. Let

$$h(x) = P_x(V_N < V_0)$$

be the happy event that our gambler reaches the goal of $\$N$ before going bankrupt when starting with $\$x$. Thanks to our definition of V_x as the minimum of $n \geq 0$ with $X_n = x$ we have $h(0) = 0$, and $h(N) = 1$. To calculate $h(x)$ for $0 < x < N$, we set $q = 1 - p$ to simplify the formulas, and consider what happens on the first step to arrive at

$$h(x) = ph(x+1) + qh(x-1) \tag{1.21}$$

To solve this we rearrange to get $p(h(x+1) - h(x)) = q(h(x) - h(x-1))$ and conclude

$$h(x+1) - h(x) = \frac{q}{p} \cdot (h(x) - h(x-1)) \tag{1.22}$$

If we set $c = h(1) - h(0)$, then (1.22) implies that for $x \geq 1$

$$h(x) - h(x-1) = c \left(\frac{q}{p}\right)^{x-1}$$

Summing from $x = 1$ to N, we have

$$1 = h(N) - h(0) = \sum_{x=1}^{N} h(x) - h(x-1) = c \sum_{x=1}^{N} \left(\frac{q}{p}\right)^{x-1}$$

Now for $\theta \neq 1$ the partial sum of the geometric series is

$$\sum_{j=0}^{N-1} \theta^j = \frac{1 - \theta^N}{1 - \theta} \tag{1.23}$$

To check this note that

$$(1 - \theta)(1 + \theta + \cdots \theta^{N-1}) = (1 + \theta + \cdots \theta^{N-1})$$

$$- (\theta + \theta^2 + \cdots \theta^N) = 1 - \theta^N$$

Using (1.23) we see that $c = (1 - \theta)/(1 - \theta^N)$ with $\theta = q/p$. Summing and using the fact that $h(0) = 0$, we have

$$h(x) = h(x) - h(0) = c \sum_{i=0}^{x-1} \theta^i = c \cdot \frac{1 - \theta^x}{1 - \theta} = \frac{1 - \theta^x}{1 - \theta^N}$$

Recalling the definition of $h(x)$ and rearranging the fraction we have

$$P_x(V_N < V_0) = \frac{\theta^x - 1}{\theta^N - 1} \qquad \text{where } \theta = \frac{1-p}{p} \tag{1.24}$$

To see what (1.24) says in a concrete example, we consider:

Example 1.45 (Roulette). If we bet \$1 on red on a roulette wheel with 18 red, 18 black, and 2 green (0 and 00) holes, we win \$1 with probability $18/38 = 0.4737$ and lose \$1 with probability $20/38$. Suppose we bring \$50 to the casino with the hope of reaching \$100 before going bankrupt. What is the probability we will succeed?

Here $\theta = q/p = 20/18$, so (1.24) implies

$$P_{50}(V_{100} < V_0) = \frac{\left(\frac{20}{18}\right)^{50} - 1}{\left(\frac{20}{18}\right)^{100} - 1}$$

Using $(20/18)^{50} = 194$, we have

$$P_{50}(V_{100} < V_0) = \frac{194 - 1}{(194)^2 - 1} = \frac{1}{194 + 1} = 0.005128$$

Now let's turn things around and look at the game from the viewpoint of the casino, i.e., $p = 20/38$. Suppose that the casino starts with the rather modest capital of $x = 100$. (1.24) implies that the probability they will reach N before going bankrupt is

$$\frac{(9/10)^{100} - 1}{(9/10)^N - 1}$$

If we let $N \to \infty$, $(9/10)^N \to 0$ so the answer converges to

$$1 - (9/10)^{100} = 1 - 2.656 \times 10^{-5}$$

If we increase the capital to \$200, then the failure probability is squared, since to become bankrupt we must first lose \$100 and then lose our second \$100. In this case the failure probability is incredibly small: $(2.656 \times 10^{-5})^2 = 7.055 \times 10^{-10}$.

From the last analysis we see that if $p > 1/2$, $q/p < 1$ and letting $N \to \infty$ in (1.24) gives

$$P_x(V_0 = \infty) = 1 - \left(\frac{q}{p}\right)^x \quad \text{and} \quad P_x(V_0 < \infty) = \left(\frac{q}{p}\right)^x. \qquad (1.25)$$

To see that the form of the last answer makes sense, note that to get from x to 0 we must go $x \to x - 1 \to x_2 \ldots \to 1 \to 0$, so

$$P_x(V_0 < \infty) = P_1(V_0 < \infty)^x.$$

Example 1.46. If we rearrange the matrix for the seven state chain in Example 1.14 we get

	2	3	1	5	4	6	7
2	.2	.3	.1	0	.4	0	0
3	0	.5	0	.2	.3	0	0
1	0	0	.7	.3	0	0	0
5	0	0	.6	.4	0	0	0
4	0	0	0	0	.5	.5	0
6	0	0	0	0	0	.2	.8
7	0	0	0	0	1	0	0

We will now use what we have learned about exit distributions to find $\lim_{n\to\infty} p^n(i,j)$.

The first step is to note that by our formula for two state chains the stationary distribution on the closed irreducible set $A = \{1,5\}$ is $2/3, 1/3$. With a little more work one concludes that the stationary distribution on $B = \{4,6,7\}$ is $8/17, 5/17, 4/17$ the third row of

$$\begin{pmatrix} .5 & -.5 & 1 \\ 0 & .8 & 1 \\ -1 & 0 & 1 \end{pmatrix}^{-1}$$

If we collapse the closed irreducible sets into a single state, then we have:

	2	3	A	B
2	.2	.3	.1	.4
3	0	.5	.2	.3

If we let r be the part of the transition probability for states 2 and 3, then

$$(I-r)^{-1}\begin{pmatrix} .1 \\ .2 \end{pmatrix} = \begin{pmatrix} 11/40 \\ 3/5 \end{pmatrix}$$

Combining these observations we see that the limit is

	2	3	1	5	4	6	7
2	0	0	11/60	11/120	29/85	29/136	29/170
3	0	0	4/15	2/15	24/85	15/85	12/85
1	0	0	2/3	1/3	0	0	0
5	0	0	2/3	1/3	0	0	0
4	0	0	0	0	8/17	5/17	4/17
6	0	0	0	0	8/17	5/17	4/17
7	0	0	0	0	8/17	5/17	4/17

1.10 Exit Times

To motivate developments we begin with an example.

Example 1.47 (Two Year College). In Example 1.40 we introduced a Markov chain with state space $1 = $ freshman, $2 = $ sophomore, $G = $ graduate, $D = $ dropout, and transition probability

	1	2	G	D
1	0.25	0.6	0	0.15
2	0	0.2	0.7	0.1
G	0	0	1	0
D	0	0	0	1

On the average how many years does a student take to graduate or drop out?

Let $g(x)$ be the expected time for a student starting in state x. $g(G) = g(D) = 0$. By considering what happens on one step

$$g(1) = 1 + 0.25g(1) + 0.6g(2)$$

$$g(2) = 1 + 0.2g(2)$$

where the $1+$ is due to the fact that after the jump has been made one year has elapsed. To solve for g, we note that the second equation implies $g(2) = 1/0.8 = 1.25$ and then the first that

$$g(1) = \frac{1 + 0.6(1.25)}{0.75} = \frac{1.75}{0.75} = 2.3333$$

Example 1.48 (Tennis). In Example 1.41 we formulated the last portion of the game as a Markov chain in which the state is the difference of the scores. The state space was $S = \{2, 1, 0, -1, -2\}$ with 2 (win for server) and -2 (win for opponent). The transition probability was

	2	1	0	-1	-2
2	1	0	0	0	0
1	.6	0	.4	0	0
0	0	.6	0	.4	0
-1	0	0	.6	0	.4
-2	0	0	0	0	1

Let $g(x)$ be the expected time to complete the game when the current state is x. By considering what happens on one step

$$g(x) = 1 + \sum_y p(x, y)g(y)$$

Since $g(2) = g(-2) = 0$, if we let $r(x, y)$ be the restriction of the transition probability to $1, 0, -1$ we have

$$g(x) - \sum_y r(x, y)g(y) = 1$$

Writing 1 for a 3×1 matrix (i.e., column vector) with all 1's we can write this as

$$(I - r)g = 1$$

so $g = (I - r)^{-1}1$.

There is another way to see this. If $N(y)$ is the number of visits to y at times $n \geq 0$, then from (1.12)

$$E_x N(y) = \sum_{n=0}^{\infty} r^n(x, y)$$

To see that this is $(I - r)^{-1}(x, y)$ note that $(I - r)(I + r + r^2 + r^3 + \cdots)$

$$= (I + r + r^2 + r^3 + \cdots) - (r + r^2 + r^3 + r^4 \cdots) = I$$

If T is the duration of the game, then $T = \sum_y N(y)$ so

$$E_x T = (I - r)^{-1}1 \tag{1.26}$$

To solve the problem now we note that

$$I - r = \begin{pmatrix} 1 & -.4 & 0 \\ -.6 & 1 & -.4 \\ 0 & -.6 & 1 \end{pmatrix} \qquad (I - r)^{-1} = \begin{pmatrix} 19/13 & 10/13 & 4/13 \\ 15/13 & 25/13 & 10/13 \\ 9/13 & 15/13 & 19/13 \end{pmatrix}$$

so $E_0 T = (15 + 25 + 10)/13 = 50/13 = 3.846$ points. Here the three terms in the sum are the expected number of visits to $-1, 0$, and 1.

Having worked two examples, it is time to show that we have computed the right answer. In some cases we will want to guess and verify the answer. In those situations it is nice to know that the solution is unique. The next result proves this.

Theorem 1.29. *Let* $V_A = \inf\{n \geq 0 : X_n \in A\}$. *Suppose* $C = S - A$ *is finite, and that* $P_x(V_A < \infty) > 0$ *for any* $x \in C$. *If* $g(a) = 0$ *for all* $a \in A$, *and for* $x \in C$ *we have*

$$g(x) = 1 + \sum_y p(x, y)g(y) \tag{1.27}$$

Then $g(x) = E_x(V_A)$.

Proof. It follows from Lemma 1.3 that $E_x V_A < \infty$ for all $x \in C$. (1.27) implies that $g(x) = 1 + E_x g(X_1)$ when $x \notin A$. The Markov property implies

$$g(x) = E_x(V_A \wedge n) + E_x g(X_{V_A \wedge n}).$$

We have to stop at time T because the equation is not valid for $x \in A$. It follows from the definition of the expected value that $E_x(V_A \wedge n) \uparrow E_x V_A$. Since g is bounded and $g(a) = 0$ for $a \in A$, we have $E_x g(X_{T \wedge n}) \to 0$. □

Example 1.49 (Waiting Time for TT). Let T_{TT} be the (random) number of times we need to flip a coin before we have gotten Tails on two consecutive tosses. To compute the expected value of T_{TT} we will introduce a Markov chain with states 0, 1, 2 = the number of Tails we have in a row.

Since getting a Tails increases the number of Tails we have in a row by 1, but getting a Heads sets the number of Tails we have in a row to 0, the transition matrix is

	0	1	2
0	1/2	1/2	0
1	1/2	0	1/2
2	0	0	1

Since we are not interested in what happens after we reach 2 we have made 2 an absorbing state. If we let $V_2 = \min\{n \geq 0 : X_n = 2\}$ and $g(x) = E_x V_2$, then one step reasoning gives

$$g(0) = 1 + .5g(0) + .5g(1)$$
$$g(1) = 1 + .5g(0)$$

Plugging the second equation into the first gives $g(0) = 1.5 + .75g(0)$, so $.25g(0) = 1.5$ or $g(0) = 6$. To do this with the previous approach we note

$$I - r = \begin{pmatrix} 1/2 & -1/2 \\ -1/2 & 1 \end{pmatrix} \qquad (I - r)^{-1} = \begin{pmatrix} 4 & 2 \\ 2 & 2 \end{pmatrix}$$

so $E_0 V_2 = 6$.

Example 1.50 (Waiting Time for HT). Let T_{HT} be the (random) number of times we need to flip a coin before we have gotten a Heads followed by a Tails. Consider X_n is Markov chain with transition probability:

$$
\begin{array}{c}
\textbf{HH HT TH TT} \\
\begin{array}{l|cccc}
\textbf{HH} & 1/2 & 1/2 & 0 & 0 \\
\textbf{HT} & 0 & 0 & 1/2 & 1/2 \\
\textbf{TH} & 1/2 & 1/2 & 0 & 0 \\
\textbf{TT} & 0 & 0 & 1/2 & 1/2
\end{array}
\end{array}
$$

If we eliminate the row and the column for HT, then

$$
I - r = \begin{pmatrix} 1/2 & 0 & 0 \\ -1/2 & 1 & 0 \\ 0 & -1/2 & 1/2 \end{pmatrix} \qquad (I-r)^{-1}\mathbf{1} = \begin{pmatrix} 2 \\ 2 \\ 4 \end{pmatrix}
$$

To compute the expected waiting time for our original problem, we note that after the first two tosses we have each of the four possibilities with probability 1/4 so

$$
ET_{HT} = 2 + \frac{1}{4}(0 + 2 + 2 + 4) = 4
$$

Why is $ET_{TT} = 6$ while $ET_{HT} = 4$? To explain we begin by noting that $E_y T_y = 1/\pi(y)$ and the stationary distribution assigns probability 1/4 to each state. One can verify this and check that convergence to equilibrium is rapid by noting that all the entries of p^2 are equal to 1/4. Our identity implies that

$$
E_{HT} T_{HT} = \frac{1}{\pi(HT)} = 4
$$

To get from this to what we wanted to calculate, note that if we start with a H at time -1 and a T at time 0, then we have nothing that will help us in the future, so the expected waiting time for a HT when we start with nothing is the same.

   When we consider TT, our identity again gives

$$
E_{TT} T_{TT} = \frac{1}{\pi(TT)} = 4
$$

However, this time if we start with a T at time -1 and a T at time 0, then a T at time 1 will give us a TT and a return to TT at time 1; while if we get a H at time 1, then we have wasted 1 turn and we have nothing that can help us later, so

$$
4 = E_{TT} T_{TT} = \frac{1}{2} \cdot 1 + \frac{1}{2} \cdot (1 + ET_{TT})
$$

Solving gives $ET_{TT} = 6$, so it takes longer to observe TT. The reason for this, which can be seen in the last equation, is that once we have one TT, we will get another one with probability 1/2, while occurrences of HT cannot overlap.

In the Exercises 1.49 and 1.50 we will consider waiting times for three coin patterns. The most interesting of these is $ET_{HTH} = ET_{THT}$.

Example 1.51. On the average how many times do we need to roll one die in order to see a run of six rolls with all different numbers? To formulate the problem we need some notation. Let $X_n, n \geq 1$ be independent and equal to 1, 2, 3, 4, 5, 6, with probability 1/6 each. Let $K_n = \max\{k : X_n, X_{n-1}, \ldots X_{n-k+1}\}$ are all different. To explain the last definition note that if the last five rolls are

$$\frac{6}{n-4} \quad \frac{2}{n-3} \quad \frac{6}{n-2} \quad \frac{5}{n-1} \quad \frac{1}{n}$$

then $K_n = 4$. K_n is a Markov chain with state space $\{1, 2, 3, 4, 5, 6\}$. To begin to compute the transition probability note that if in the example drawn above $X_{n+1} \in \{3, 4\}$ then $K_{n+1} = 5$ while if $X_{n+1} = 1, 5, 6, 2$ then $K_{n+1} = 1, 2, 3, 4$. Extending the reasoning we see that the transition probability is

	1	2	3	4	5	6
1	1/6	5/6	0	0	0	0
2	1/6	1/6	4/6	0	0	0
3	1/6	1/6	1/6	3/6	0	0
4	1/6	1/6	1/6	1/6	2/6	0
5	1/6	1/6	1/6	1/6	1/6	1/6
6	1/6	1/6	1/6	1/6	1/6	1/6

If we let r be the portion of the transition probability for states $\{1, 2, 3, 4, 5\}$, then

$$(I - r)^{-1} = \begin{pmatrix} 14.7 & 23.5 & 23 & 15 & 6 \\ 13.5 & 23.5 & 23 & 15 & 6 \\ 13.2 & 22 & 23 & 15 & 6 \\ 12.6 & 21 & 21 & 15 & 6 \\ 10.8 & 18 & 18 & 12 & 6 \end{pmatrix} \tag{1.28}$$

To see why the last column consists of 6's recall that multiplying $(I - r)^{-1}$ by the column vector $(0, 0, 0, 0, 1/6)$ gives the probability we exist at 6 which is 1. In next to last column $E_x N(5) = 15$ for $x = 1, 2, 3, 4$, since in each case we will wander around for a while before we hit 5.

Using (1.28) the expected exit times are

$$(I - r)^{-1}\mathbf{1} = \begin{pmatrix} 82.2 \\ 81 \\ 79.2 \\ 75.6 \\ 64.8 \end{pmatrix}$$

To check the order of magnitude of the answer we note that the probability six rolls of a die will be different is

$$\frac{6!}{6^6} = .015432 = \frac{1}{64.8}$$

To connect this with the last entry note that

$$E_5 T_6 = E_6 T_6 = \frac{1}{\pi(6)}$$

Example 1.52 (Duration of Fair Games). Consider the gambler's ruin chain in which $p(i, i+1) = p(i, i-1) = 1/2$. Let $\tau = \min\{n : X_n \notin (0, N)\}$. We claim that

$$E_x \tau = x(N - x) \tag{1.29}$$

To see what formula (1.29) says, consider matching pennies. There $N = 25$ and $x = 15$, so the game will take $15 \cdot 10 = 150$ flips on the average. If there are twice as many coins, $N = 50$ and $x = 30$, then the game takes $30 \cdot 20 = 600$ flips on the average, or four times as long.

There are two ways to prove this.

Verify the Guess Let $g(x) = x(N - x)$. Clearly, $g(0) = g(N) = 0$. If $0 < x < N$, then by considering what happens on the first step we have

$$g(x) = 1 + \frac{1}{2}g(x + 1) + \frac{1}{2}g(x - 1)$$

If $g(x) = x(N - x)$, then the right-hand side is

$$= 1 + \frac{1}{2}(x + 1)(N - x - 1) + \frac{1}{2}(x - 1)(N - x + 1)$$

$$= 1 + \frac{1}{2}[x(N - x) - x + N - x - 1] + \frac{1}{2}[x(N - x) + x - (N - x + 1)]$$

$$= 1 + x(N - x) - 1 = x(N - x)$$

Derive the Answer (1.27) implies that

$$g(x) = 1 + (1/2)g(x + 1) + (1/2)g(x - 1)$$

Rearranging gives

$$g(x + 1) - g(x) = -2 + g(x) - g(x - 1)$$

Setting $g(1) - g(0) = c$ we have $g(2) - g(1) = c - 2$, $g(3) - g(2) = c - 4$ and in general that

$$g(k) - g(k-1) = c - 2(k-1)$$

Using $g(0) = 0$ and summing we have

$$0 = g(N) = \sum_{k=1}^{N} c - 2(k-1) = cN - 2 \cdot \frac{N(N-1)}{2}$$

since, as one can easily check by induction, $\sum_{j=1}^{m} j = m(m+1)/2$. Solving gives $c = (N-1)$. Summing again, we see that

$$g(x) = \sum_{k=1}^{x} (N-1) - 2(k-1) = x(N-1) - x(x+1) = x(N-x)$$

Example 1.53 (Duration of Nonfair Games). Consider the gambler's ruin chain in which $p(i, i+1)p$ and $p(i, i-1) = q$, where $p \neq q$. Let $\tau = \min\{n : X_n \notin (0, N)\}$. We claim that

$$E_x\tau = \frac{x}{q-p} - \frac{N}{q-p} \cdot \frac{1 - (q/p)^x}{1 - (q/p)^N} \tag{1.30}$$

This time the derivation is somewhat tedious, so we will just verify the guess. We want to show that $g(x) = 1 + pg(x+1) + qg(x-1)$. Plugging the formula into the right-hand side:

$$= 1 + p\frac{x+1}{q-p} + q\frac{x-1}{q-p} - \frac{N}{q-p}\left[p \cdot \frac{1 - (q/p)^{x+1}}{1 - (q/p)^N} + q\frac{1 - (q/p)^{x-1}}{1 - (q/p)^N} \right]$$

$$= 1 + \frac{x}{q-p} + \frac{p-q}{q-p} - \frac{N}{q-p}\left[\frac{p + q - (q/p)^x(q+p)}{1 - (q/p)^N} \right]$$

which $= g(x)$ since $p + q = 1$.

To see what this says note that if $p < q$ then $q/p > 1$ so

$$\frac{N}{1 - (q/p)^N} \to 0 \quad \text{and} \quad g(x) = \frac{x}{q-p} \tag{1.31}$$

To see this is reasonable note that our expected value on one play is $p - q$, so we lose an average of $q - p$ per play, and it should take an average of $x/(q-p)$ to lose x dollars.

When $p > q$, $(q/p)^N \to 0$, so doing some algebra

$$g(x) \approx \frac{N-x}{p-q}[1-(q/p)^x] + \frac{x}{p-q}(q/p)^x$$

Using (1.25) we see that the probability of not hitting 0 is $1 - (q/p)^x$. In this case, since our expected winnings per play is $p - q$, it should take about $(N - x)/(p - q)$ plays to get to N. The second term represents the contribution to the expected value from paths that end at 0, but it is hard to explain why the term has exactly this form.

1.11   Infinite State Spaces*

In this section we consider chains with an infinite state space. The major new complication is that recurrence is not enough to guarantee the existence of a stationary distribution.

Example 1.54 (Reflecting Random Walk). Imagine a particle that moves on $\{0, 1, 2, \ldots\}$ according to the following rules. It takes a step to the right with probability p. It attempts to take a step to the left with probability $1 - p$, but if it is at 0 and tries to jump to the left, it stays at 0, since there is no -1 to jump to. In symbols,

$$p(i, i + 1) = p \quad \text{when } i \geq 0$$
$$p(i, i - 1) = 1 - p \quad \text{when } i \geq 1$$
$$p(0, 0) = 1 - p$$

This is a birth and death chain, so we can solve for the stationary distribution using the detailed balance equations:

$$p\pi(i) = (1 - p)\pi(i + 1) \quad \text{when } i \geq 0$$

Rewriting this as $\pi(i + 1) = \pi(i) \cdot p/(1 - p)$ and setting $\pi(0) = c$, we have

$$\pi(i) = c\left(\frac{p}{1-p}\right)^i \tag{1.32}$$

There are now three cases to consider:

$p < 1/2$: $p/(1 - p) < 1$. $\pi(i)$ decreases exponentially fast, so $\sum_i \pi(i) < \infty$, and we can pick c to make π a stationary distribution. To find the value of c to make π a probability distribution we recall

$$\sum_{i=0}^{\infty} \theta^i = 1/(1 - \theta) \quad \text{when } \theta < 1.$$

Taking $\theta = p/(1-p)$ and hence $1 - \theta = (1-2p)/(1-p)$, we see that the sum of the $\pi(i)$ defined in $(*)$ is $c(1-p)/(1-2p)$, so

$$\pi(i) = \frac{1-2p}{1-p} \cdot \left(\frac{p}{1-p}\right)^i = (1-\theta)\theta^i \qquad (1.33)$$

To confirm that we have succeeded in making the $\pi(i)$ add up to 1, note that if we are flipping a coin with a probability θ of Heads, then the probability of getting i Heads before we get our first Tails is given by $\pi(i)$.

The reflecting random walk is clearly irreducible. To check that it is aperiodic note that $p(0,0) > 0$ implies 0 has period 1, and then Lemma 1.17 implies that all states have period 1. Using the convergence theorem, Theorem 1.19, now we see that

I. When $p < 1/2$, $P(X_n = j) \to \pi(j)$, the stationary distribution in (1.33).

Using Theorem 1.21 now,

$$E_0 T_0 = \frac{1}{\pi(0)} = \frac{1}{1-\theta} = \frac{1-p}{1-2p} \qquad (1.34)$$

It should not be surprising that the system stabilizes when $p < 1/2$. In this case movements to the left have a higher probability than to the right, so there is a drift back toward 0. On the other hand, if steps to the right are more frequent than those to the left, then the chain will drift to the right and wander off to ∞.

II. When $p > 1/2$ all states are transient.

(1.25) implies that if $x > 0$, $P_x(T_0 < \infty) = ((1-p)/p)^x$.

To figure out what happens in the borderline case $p = 1/2$, we use results from Sects. 1.8 and 1.9. Recall we have defined $V_y = \min\{n \geq 0 : X_n = y\}$ and (1.19) tells us that if $x > 0$

$$P_x(V_N < V_0) = x/N$$

If we keep x fixed and let $N \to \infty$, then $P_x(V_N < V_0) \to 0$ and hence

$$P_x(V_0 < \infty) = 1$$

In words, for any starting point x, the random walk will return to 0 with probability 1. To compute the mean return time, we note that if $\tau_N = \min\{n : X_n \notin (0, N)\}$, then we have $\tau_N \leq V_0$ and by (1.29) we have $E_1\tau_N = N - 1$. Letting $N \to \infty$ and combining the last two facts shows $E_1 V_0 = \infty$. Reintroducing our old hitting time $T_0 = \min\{n > 0 : X_n = 0\}$ and noting that on our first step we go to 0 or to 1 with probability 1/2 shows that

$$E_0 T_0 = (1/2) \cdot 1 + (1/2)E_1 V_0 = \infty$$

Summarizing the last two paragraphs, we have

III. *When* $p = 1/2$, $P_0(T_0 < \infty) = 1$ *but* $E_0 T_0 = \infty$.

Thus when $p = 1/2$, 0 is recurrent in the sense we will certainly return, but it is not recurrent in the following sense:

x is said to be **positive recurrent** if $E_x T_x < \infty$.

If a state is recurrent but not positive recurrent, i.e., $P_x(T_x < \infty) = 1$ but $E_x T_x = \infty$, then we say that x is **null recurrent**.

In our new terminology, our results for reflecting random walk say

If $p < 1/2$, 0 is positive recurrent
If $p = 1/2$, 0 is null recurrent
If $p > 1/2$, 0 is transient

In reflecting random walk, null recurrence thus represents the borderline between recurrence and transience. This is what we think in general when we hear the term. To see the reason we might be interested in positive recurrence recall that by Theorem 1.21

$$\pi(x) = \frac{1}{E_x T_x}$$

If $E_x T_x = \infty$, then this gives $\pi(x) = 0$. This observation motivates

Theorem 1.30. *For an irreducible chain the following are equivalent:*

  (i) Some state is positive recurrent.
  (ii) There is a stationary distribution π.
(iii) All states are positive recurrent.

Proof. The stationary measure constructed in Theorem 1.24 has total mass

$$\sum_y \mu(y) = \sum_{n=0}^{\infty} \sum_y P_x(X_n = y, T_x > n)$$

$$= \sum_{n=0}^{\infty} P_x(T_x > n) = E_x T_x$$

so (i) implies (ii). Noting that irreducibility implies $\pi(y) > 0$ for all y and then using $\pi(y) = 1/E_y T_y$ shows that (ii) implies (iii). It is trivial that (iii) implies (i). $\square$

Our next example may at first seem to be quite different. In a branching process 0 is an absorbing state, so by Theorem 1.5 all the other states are transient. However, as the story unfolds we will see that branching processes have the same trichotomy as random walks do.

Example 1.55 (Branching Processes). Consider a population in which each individual in the nth generation gives birth to an independent and identically distributed number of children. The number of individuals at time n, X_n is a Markov chain with transition probability given in Example 1.8. As announced there, we are interested in the question:

Q. What is the probability the species avoids extinction?

Here "extinction" means becoming absorbed state at 0. As we will now explain, whether this is possible or not can be determined by looking at the average number of offspring of one individual:

$$\mu = \sum_{k=0}^{\infty} k p_k$$

If there are m individuals at time $n-1$, then the mean number at time n is $m\mu$. More formally the conditional expectation given X_{n-1}

$$E(X_n|X_{n-1}) = \mu X_{n-1}$$

Taking expected values of both sides gives $EX_n = \mu EX_{n-1}$. Iterating gives

$$EX_n = \mu^n EX_0 \qquad (1.35)$$

If $\mu < 1$, then $EX_n \to 0$ exponentially fast. Using the inequality

$$EX_n \geq P(X_n \geq 1)$$

it follows that $P(X_n \geq 1) \to 0$ and we have

I. *If $\mu < 1$, then extinction occurs with probability 1.*

To treat the cases $\mu \geq 1$ we will use a one-step calculation. Let ρ be the probability that this process dies out (i.e., reaches the absorbing state 0) starting from $X_0 = 1$. If there are k children in the first generation, then in order for extinction to occur, the family line of each child must die out, an event of probability ρ^k, so we can reason that

$$\rho = \sum_{k=0}^{\infty} p_k \rho^k \qquad (1.36)$$

If we let $\phi(\theta) = \sum_{k=0}^{\infty} p_k \theta^k$ be the generating function of the distribution p_k, then the last equation can be written simply as $\rho = \phi(\rho)$ (Fig. 1.4).

The equation in (1.36) has a trivial root at $\rho = 1$ since $\phi(\rho) = \sum_{k=0}^{\infty} p_k \rho^k = 1$. The next result identifies the root that we want:

Lemma 1.31. *The extinction probability ρ is the smallest solution of the equation $\phi(x) = x$ with $0 \leq x \leq 1$.*

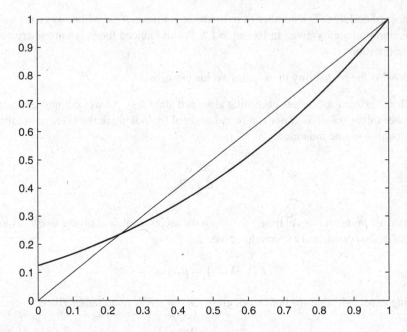

Fig. 1.4 Generating function for Binomial(3,1/2)

Proof. Extending the reasoning for (1.36) we see that in order for the process to hit 0 by time n, all of the processes started by first-generation individuals must hit 0 by time $n-1$, so

$$P(X_n = 0) = \sum_{k=0}^{\infty} p_k P(X_{n-1} = 0)^k$$

From this we see that if $\rho_n = P(X_n = 0)$ for $n \geq 0$, then $\rho_n = \phi(\rho_{n-1})$ for $n \geq 1$.

Since 0 is an absorbing state, $\rho_0 \leq \rho_1 \leq \rho_2 \leq \ldots$ and the sequence converges to a limit ρ_∞. Letting $n \to \infty$ in $\rho_n = \phi(\rho_{n-1})$ implies that $\rho_\infty = \phi(\rho_\infty)$, i.e., ρ_∞ is a solution of $\phi(x) = x$. To complete the proof now let ρ be the smallest solution. Clearly $\rho_0 = 0 \leq \rho$. Using lthe fact that ϕ is increasing, it follows that $\rho_1 = \phi(\rho_0) \leq \phi(\rho) = \rho$. Repeating the argument we have $\rho_2 \leq \rho$, $\rho_3 \leq \rho$ and so on. Taking limits we have $\rho_\infty \leq \rho$. However, ρ is the smallest solution, so we must have $\rho_\infty = \rho$.   □

To see what this says, let us consider a concrete example.

Example 1.56 (Binary Branching). Suppose $p_2 = a$, $p_0 = 1 - a$, and the other $p_k = 0$. In this case $\phi(\theta) = a\theta^2 + 1 - a$, so $\phi(x) = x$ means

$$0 = ax^2 - x + 1 - a = (x-1)(ax - (1-a))$$

The roots are 1 and $(1 - a)/a$. If $a \le 1/2$, then the smallest root is 1, while if $a > 1/2$ the smallest root is $(1 - a)/a$.

Noting that $a \le 1/2$ corresponds to mean $\mu \le 1$ in binary branching motivates the following guess:

II. *If $\mu > 1$, then there is positive probability of avoiding extinction.*

Proof. In view of Lemma 1.31, we only have to show there is a root < 1. We begin by discarding a trivial case. If $p_0 = 0$, then $\phi(0) = 0$, 0 is the smallest root, and there is no probability of dying out. If $p_0 > 0$, then $\phi(0) = p_0 > 0$. Differentiating the definition of ϕ, we have

$$\phi'(x) = \sum_{k=1}^{\infty} p_k \cdot k x^{k-1} \quad \text{so} \quad \phi'(1) = \sum_{k=1}^{\infty} k p_k = \mu$$

If $\mu > 1$, then the slope of ϕ at $x = 1$ is larger than 1, so if ϵ is small, then $\phi(1 - \epsilon) < 1 - \epsilon$. Combining this with $\phi(0) > 0$ we see there must be a solution of $\phi(x) = x$ between 0 and $1 - \epsilon$. See the figure in the proof of (7.6). $\quad\square$

Turning to the borderline case:

III. *If $\mu = 1$ and we exclude the trivial case $p_1 = 1$, then extinction occurs with probability 1.*

Proof. By Lemma 1.31 we only have to show that there is no root < 1. To do this we note that if $p_1 < 1$, then for $y < 1$

$$\phi'(x) = \sum_{k=1}^{\infty} p_k \cdot k x^{k-1} < \sum_{k=1}^{\infty} p_k k = 1$$

so if $x < 1$ then $\phi(x) = \phi(1) - \int_x^1 \phi'(y)\, dy > 1 - (1 - x) = x$. Thus $\phi(x) > x$ for all $x < 1$. $\quad\square$

Note that in binary branching with $a = 1/2$, $\phi(x) = (1 + x^2)/2$, so if we try to solve $\phi(x) = x$ we get

$$0 = 1 - 2x + x^2 = (1 - x)^2$$

i.e., a double root at $x = 1$. In general when $\mu = 1$, the graph of ϕ is tangent to the diagonal (x, x) at $x = 1$. This slows down the convergence of ρ_n to 1 so that it no longer occurs exponentially fast.

In more advanced treatments, it is shown that if the offspring distribution has mean 1 and variance $\sigma^2 > 0$, then

$$P_1(X_n > 0) \sim \frac{2}{n\sigma^2}$$

This is not easy even for the case of binary branching, so we refer to reader to Sect. 1.9 of Athreya and Ney (1972) for a proof. We mention the result here because it allows us to see that the expected time for the process to die out $\sum_n P_1(T_0 > n) = \infty$. If we modify the branching process, so that $p(0, 1) = 1$ then in the modified process

If $\mu < 1, 0$ is positive recurrent
If $\mu = 1, 0$ is null recurrent
If $\mu > 1, 0$ is transient

1.12   Chapter Summary

A Markov chain with transition probability p is defined by the property that given the present state the rest of the past is irrelevant for predicting the future:

$$P(X_{n+1} = y | X_n = x, X_{n-1} = x_{n-1}, \ldots, X_0 = x_0) = p(x, y)$$

The m step transition probability

$$p^m(x, y) = P(X_{n+m} = y | X_n = x)$$

is the mth power of the matrix p.

Recurrence and Transience

The first thing we need to determine about a Markov chain is which states are recurrent and which are transient. To do this we let $T_y = \min\{n \geq 1 : X_n = y\}$ and let

$$\rho_{xy} = P_x(T_y < \infty)$$

When $x \neq y$ this is the probability X_n ever visits y starting at x. When $x = y$ this is the probability X_n returns to y when it starts at y. We restrict to times $n \geq 1$ in the definition of T_y so that we can say: y is recurrent if $\rho_{yy} = 1$ and transient if $\rho_{yy} < 1$.
   Transient states in a finite state space can all be identified using

Theorem 1.5. *If $\rho_{xy} > 0$, but $\rho_{yx} < 1$, then x is transient.*

   Once the transient states are removed we can use

Theorem 1.7. *If C is a finite closed and irreducible set, then all states in C are recurrent.*

Here A is closed if $x \in A$ and $y \notin A$ implies $p(x, y) = 0$, and B is irreducible if $x, y \in B$ implies $\rho_{xy} > 0$.

The keys to the proof of Theorem 1.7 are: (i) If x is recurrent and $\rho_{xy} > 0$, then y is recurrent, and (ii) In a finite closed set there has to be at least one recurrent state. To prove these results, it was useful to know that if $N(y)$ is the number of visits to y at times $n \geq 1$ then

$$\sum_{n=1}^{\infty} p^n(x, y) = E_x N(y) = \frac{\rho_{xy}}{1 - \rho_{yy}}$$

so y is recurrent if and only if $E_y N(y) = \infty$.

Theorems 1.5 and 1.7 allow us to decompose the state space and simplify the study of Markov chains.

Theorem 1.8. *If the state space S is finite, then S can be written as a disjoint union $T \cup R_1 \cup \cdots \cup R_k$, where T is a set of transient states and the R_i, $1 \leq i \leq k$, are closed irreducible sets of recurrent states.*

Stationary Distributions

A stationary measure is a nonnegative solution of $\mu p = \mu$. A stationary distribution is a nonnegative solution of $\pi p = \pi$ normalized so that the entries sum to 1. The first question is: do these things exist?

Theorem 1.24. *Suppose p is irreducible and recurrent. Let $x \in S$ and let $T_x = \inf\{n \geq 1 : X_n = x\}$.*

$$\mu_x(y) = \sum_{n=0}^{\infty} P_x(X_n = y, T_x > n)$$

defines a stationary measure with $0 < \mu_x(y) < \infty$ for all y.

If the state space S is finite and irreducible, there is a unique stationary distribution. More generally if $E_x T_x < \infty$, i.e., x is positive recurrent, then $\mu_x(y)/E_x T_x$ is a stationary distribution. Since $\mu_x(x) = 1$ we see that

$$\pi(x) = \frac{1}{E_x T_x}$$

If there are k states, then the stationary distribution π can be computed by the following procedure. Form a matrix A by taking the first $k - 1$ columns of $p - I$ and adding a final column of 1's. The equations $\pi p = \pi$ and $\pi_1 + \cdots \pi_k = 1$ are equivalent to

$$\pi A = \begin{pmatrix} 0 \ldots 0 & 1 \end{pmatrix}$$

so we have

$$\pi = \begin{pmatrix} 0 \dots 0 \ 1 \end{pmatrix} A^{-1}$$

or π is the bottom row of A^{-1}.

In two situations, the stationary distribution is easy to compute. (i) If the chain is doubly stochastic, i.e., $\sum_x p(x,y) = 1$, and has k states, then the stationary distribution is uniform $\pi(x) = 1/k$. (ii) π is a stationary distribution if the detailed balance condition holds

$$\pi(x)p(x,y) = \pi(y)p(y,x)$$

Birth and death chains, defined by the condition that $p(x,y) = 0$ if $|x-y| > 1$ always have stationary distributions with this property. If the state space is $\ell, \ell+1, \dots r$ then π can be found by setting $\pi(\ell) = c$, solving for $\pi(x)$ for $\ell < x \leq r$, and then choosing c to make the probabilities sum to 1.

Convergence Theorems

Transient states y have $p^n(x,y) \to 0$, so to investigate the convergence of $p^n(x,y)$ it is enough, by the decomposition theorem, to suppose the chain is irreducible and all states are recurrent. The period of a state is the greatest common divisor of $I_x = \{n \geq 1 : p^n(x,x) > 0\}$. If the period is 1, x is said to be aperiodic. A simple sufficient condition to be aperiodic is that $p(x,x) > 0$. To compute the period it is useful to note that if $\rho_{xy} > 0$ and $\rho_{yx} > 0$ then x and y have the same period. In particular all of the states in an irreducible set have the same period.

The three main results about the asymptotic behavior of Markov chains are

Theorem 1.19. *Suppose p is irreducible, aperiodic, and has a stationary distribution π. Then as $n \to \infty$, $p^n(x,y) \to \pi(y)$.*

Theorem 1.20. *Suppose p is irreducible and recurrent. If $N_n(y)$ be the number of visits to y up to time n, then*

$$\frac{N_n(y)}{n} \to \frac{1}{E_y T_y}$$

Theorem 1.22. *Suppose p is irreducible, has stationary distribution π, and $\sum_x |f(x)|\pi(x) < \infty$ then*

$$\frac{1}{n}\sum_{m=1}^{n} f(X_m) \to \sum_x f(x)\pi(x)$$

Exit Distributions

If F is a subset of S let $V_F = \min\{n \geq 0 : X_n \in F\}$.

Theorem 1.28. *Consider a Markov chain with finite state space S. Let $A, B \subset S$, so that $C = S - \{a, b\}$ is finite, and $P_x(V_A \wedge V_B < \infty) > 0$ for all $x \in C$. If $h(a) = 1$ for $a \in A$, $h(b) = 0$ for $b \in B$, and*

$$h(x) = \sum_y p(x, y)h(y) \quad \text{for } x \in C$$

then $h(x) = P_x(V_A < V_B)$.

Let $r(x, y)$ be the part of the matrix $p(x, y)$ with $x, y \in C$ and for $x \in C$ let $v(x) = \sum_{y \in A} p(x, y)$ which we think of as a column vector. Since $h(a) = 1$ for $a \in A$ and $h(b) = 0$ for $b \in B$, the equation for h can be written for $x \in C$ as

$$h(x) = v(x) + \sum_y r(x, y)h(y)$$

and the solution is

$$h = (I - r)^{-1}v.$$

Exit Times

Theorem 1.29. *Consider a Markov chain with finite state space S. Let $A \subset S$ so that $C = S - A$ is finite and $P_x(V_A < \infty) > 0$ for $x \in C$. If $g(a) = 0$ for all $a \in A$, and*

$$g(x) = 1 + \sum_y p(x, y)g(y) \quad \text{for } x \in C$$

then $g(x) = E_x(V_A)$.

Since $g(x) = 0$ for $x \in A$ the equation for g can be written for $x \in C$ as

$$g(x) = 1 + \sum_y r(x, y)g(y)$$

so if we let **1** be a column vector consisting of all 1's then the last equation says $(I - r)g = \mathbf{1}$ and the solution is

$$g = (I - r)^{-1}\mathbf{1}.$$

$(I - r)^{-1}(x, y)$ is the expected number of visits to y starting from x. When we multiply by $\mathbf{1}$ we sum over all $y \in C$, so the result is the expected exit time.

1.13   Exercises

Understanding the Definitions

1.1. A fair coin is tossed repeatedly with results $Y_0, Y_1, Y_2, \ldots$ that are 0 or 1 with probability 1/2 each. For $n \geq 1$ let $X_n = Y_n + Y_{n-1}$ be the number of 1's in the $(n-1)$th and nth tosses. Is X_n a Markov chain?

1.2. Five white balls and five black balls are distributed in two urns in such a way that each urn contains five balls. At each step we draw one ball from each urn and exchange them. Let X_n be the number of white balls in the left urn at time n. Compute the transition probability for X_n.

1.3. We repeated roll two four sided dice with numbers 1, 2, 3, and 4 on them. Let Y_k be the sum on the kth roll, $S_n = Y_1 + \cdots + Y_n$ be the total of the first n rolls, and $X_n = S_n \pmod{6}$. Find the transition probability for X_n.

1.4. The 1990 census showed that 36 % of the households in the District of Columbia were homeowners while the remainder were renters. During the next decade 6 % of the homeowners became renters and 12 % of the renters became homeowners. What percentage were homeowners in 2000? in 2010?

1.5. Consider a gambler's ruin chain with $N = 4$. That is, if $1 \leq i \leq 3, p(i, i+1) = 0.4$, and $p(i, i-1) = 0.6$, but the endpoints are absorbing states: $p(0,0) = 1$ and $p(4, 4) = 1$ Compute $p^3(1, 4)$ and $p^3(1, 0)$.

1.6. A taxicab driver moves between the airport A and two hotels B and C according to the following rules. If he is at the airport, he will be at one of the two hotels next with equal probability. If at a hotel, then he returns to the airport with probability 3/4 and goes to the other hotel with probability 1/4. (a) Find the transition matrix for the chain. (b) Suppose the driver begins at the airport at time 0. Find the probability for each of his three possible locations at time 2 and the probability he is at hotel B at time 3.

1.7. Suppose that the probability it rains today is 0.3 if neither of the last two days was rainy, but 0.6 if at least one of the last two days was rainy. Let the weather on day n, W_n, be R for rain, or S for sun. W_n is not a Markov chain, but the weather for the last two days $X_n = (W_{n-1}, W_n)$ is a Markov chain with four states $\{RR, RS, SR, SS\}$. (a) Compute its transition probability. (b) Compute the two-step transition probability. (c) What is the probability it will rain on Wednesday given that it did not rain on Sunday or Monday.

1.8. Consider the following transition matrices. Identify the transient and recurrent states, and the irreducible closed sets in the Markov chains. Give reasons for your answers.

(a)

	1	2	3	4	5
1	.4	.3	.3	0	0
2	0	.5	0	.5	0
3	.5	0	.5	0	0
4	0	.5	0	.5	0
5	0	.3	0	.3	.4

(b)

	1	2	3	4	5	6
1	.1	0	0	.4	.5	0
2	.1	.2	.2	0	.5	0
3	0	.1	.3	0	0	.6
4	.1	0	0	.9	0	0
5	0	0	0	.4	0	.6
6	0	0	0	0	.5	.5

(c)

	1	2	3	4	5
1	0	0	0	0	1
2	0	.2	0	.8	0
3	.1	.2	.4	.3	0
4	0	.4	0	.6	0
5	.2	0	0	0	.8

(d)

	1	2	3	4	5	6
1	.8	0	0	.2	0	0
2	0	.5	0	0	.5	0
3	0	0	.3	.4	.1	.2
4	.1	0	0	.9	0	0
5	0	.2	0	0	.8	0
6	0	.3	0	.3	0	.4

(e)

	1	2	3	4	5
1	1	0	0	0	0
2	0	2/3	0	1/3	0
3	1/8	1/4	5/8	0	0
4	0	1/6	0	5/6	0
5	1/3	0	1/3	0	1/3

1.9. Find the stationary distributions for the Markov chains with transition matrices:

(a)

	1	2	3
1	.5	.4	.1
2	.2	.5	.3
3	.1	.3	.6

(b)

	1	2	3
1	.5	.4	.1
2	.3	.4	.3
3	.2	.2	.6

(c)

	1	2	3
1	.6	.4	0
2	.2	.4	.2
3	0	.2	.8

1.10. Find the stationary distributions for the Markov chains on $\{1, 2, 3, 4\}$ with transition matrices:

(a)
$$\begin{pmatrix} .7 & 0 & .3 & 0 \\ .6 & 0 & .4 & 0 \\ 0 & .5 & 0 & .5 \\ 0 & .4 & 0 & .6 \end{pmatrix}$$

(b)
$$\begin{pmatrix} .7 & .3 & 0 & 0 \\ .2 & .5 & .3 & 0 \\ .0 & .3 & .6 & .1 \\ 0 & 0 & .2 & .8 \end{pmatrix}$$

(c)
$$\begin{pmatrix} .7 & 0 & .3 & 0 \\ .2 & .5 & .3 & 0 \\ .1 & .2 & .4 & .3 \\ 0 & .3 & 0 & .7 \end{pmatrix}$$

1.11. (a) Find the stationary distribution for the transition probability

$$
\begin{array}{c|ccc}
 & 1 & 2 & 3 \\
1 & .2 & .4 & .4 \\
2 & .1 & .6 & .3 \\
3 & .2 & .6 & .2 \\
\end{array}
$$

(b) Does it satisfy the detailed balance condition (1.11)?

1.12. Find the stationary distributions for the chains in exercises (a) 1.2, (b) 1.3, and (c) 1.7.

1.13. Consider the Markov chain with transition matrix:

$$
\begin{array}{c|cccc}
 & 1 & 2 & 3 & 4 \\
1 & 0 & 0 & 0.1 & 0.9 \\
2 & 0 & 0 & 0.6 & 0.4 \\
3 & 0.8 & 0.2 & 0 & 0 \\
4 & 0.4 & 0.6 & 0 & 0 \\
\end{array}
$$

(a) Compute p^2. (b) Find the stationary distributions of p and all of the stationary distributions of p^2. (c) Find the limit of $p^{2n}(x,x)$ as $n \to \infty$.

1.14. Do the following Markov chains converge to equilibrium?

$$
\begin{array}{c|cccc}
(a) & 1 & 2 & 3 & 4 \\
1 & 0 & 0 & 1 & 0 \\
2 & 0 & 0 & .5 & .5 \\
3 & .3 & .7 & 0 & 0 \\
4 & 1 & 0 & 0 & 0 \\
\end{array}
\qquad
\begin{array}{c|cccc}
(b) & 1 & 2 & 3 & 4 \\
1 & 0 & 1 & .0 & 0 \\
2 & 0 & 0 & 0 & 1 \\
3 & 1 & 0 & 0 & 0 \\
4 & 1/3 & 0 & 2/3 & 0 \\
\end{array}
$$

$$
\begin{array}{c|cccccc}
(c) & 1 & 2 & 3 & 4 & 5 & 6 \\
1 & 0 & .5 & .5 & 0 & 0 & 0 \\
2 & 0 & 0 & 0 & 1 & 0 & 0 \\
3 & 0 & 0 & 0 & .4 & 0 & .6 \\
4 & 1 & 0 & 0 & 0 & 0 & 0 \\
5 & 0 & 1 & 0 & 0 & 0 & 0 \\
6 & .2 & 0 & 0 & 0 & .8 & 0 \\
\end{array}
\qquad
\begin{array}{c|cccccc}
(d) & 1 & 2 & 3 & 4 & 5 & 6 \\
1 & 0 & 0 & 1 & 0 & 0 & 0 \\
2 & 1 & 0 & 0 & 0 & 0 & 0 \\
3 & 0 & .5 & 0 & 0 & .5 & 0 \\
4 & 0 & .5 & 0 & 0 & .5 & 0 \\
5 & 0 & 0 & 0 & 0 & 0 & 1 \\
6 & 0 & 0 & 0 & 1 & 0 & 0 \\
\end{array}
$$

1.15. Find $\lim_{n\to\infty} p^n(i,j)$ for the chains in parts (c), (d), and (e) of Problem 1.8.

Two State Markov Chains

1.16. Market research suggests that in a five year period 8 % of people with cable television will get rid of it, and 26 % of those without it will sign up for it. Compare the predictions of the Markov chain model with the following data on the fraction of people with cable TV: 56.4 % in 1990, 63.4 % in 1995, and 68.0 % in 2000. What is the long run fraction of people with cable TV?

1.17. A sociology professor postulates that in each decade 8 % of women in the work force leave it and 20 % of the women not in it begin to work. Compare the predictions of his model with the following data on the percentage of women working: 43.3 % in 1970, 51.5 % in 1980, 57.5 % in 1990, and 59.8 % in 2000. In the long run what fraction of women will be working?

1.18. A rapid transit system has just started operating. In the first month of operation, it was found that 25 % of commuters are using the system while 75 % are travelling by automobile. Suppose that each month 10 % of transit users go back to using their cars, while 30 % of automobile users switch to the transit system. (a) Compute the three step transition probability p^3. (b) What will be the fractions using rapid transit in the fourth month? (c) In the long run?

1.19. A regional health study indicates that from one year to the next, 75 % percent of smokers will continue to smoke while 25 % will quit. 8 % of those who stopped smoking will resume smoking while 92 % will not. If 70 % of the population were smokers in 1995, what fraction will be smokers in 1998? in 2005? in the long run?

1.20. Three of every four trucks on the road are followed by a car, while only one of every five cars is followed by a truck. What fraction of vehicles on the road are trucks?

1.21. In a test paper the questions are arranged so that 3/4's of the time a True answer is followed by a True, while 2/3's of the time a False answer is followed by a False. You are confronted with a 100 question test paper. Approximately what fraction of the answers will be True.

1.22. In unprofitable times corporations sometimes suspend dividend payments. Suppose that after a dividend has been paid the next one will be paid with probability 0.9, while after a dividend is suspended the next one will be suspended with probability 0.6. In the long run what is the fraction of dividends that will be paid?

1.23. Census results reveal that in the USA 80 % of the daughters of working women work and that 30 % of the daughters of nonworking women work. (a) Write the transition probability for this model. (b) In the long run what fraction of women will be working?

1.24. When a basketball player makes a shot then he tries a harder shot the next time and hits (H) with probability 0.4, misses (M) with probability 0.6. When he misses he is more conservative the next time and hits (H) with probability 0.7, misses (M) with probability 0.3. (a) Write the transition probability for the two state Markov chain with state space $\{H, M\}$. (b) Find the long-run fraction of time he hits a shot.

1.25. Folk wisdom holds that in Ithaca in the summer it rains 1/3 of the time, but a rainy day is followed by a second one with probability 1/2. Suppose that Ithaca weather is a Markov chain. What is its transition probability?

Chains with Three or More States

1.26. (a) Suppose brands A and B have consumer loyalties of .7 and .8, meaning that a customer who buys A one week will with probability .7 buy it again the next week, or try the other brand with .3. What is the limiting market share for each of these products? (b) Suppose now there is a third brand with loyalty .9, and that a consumer who changes brands picks one of the other two at random. What is the new limiting market share for these three products?

1.27. A midwestern university has three types of health plans: a health maintenance organization (*HMO*), a preferred provider organization (*PPO*), and a traditional fee for service plan (*FFS*). Experience dictates that people change plans according to the following transition matrix

	HMO	PPO	FFS
HMO	.85	.1	.05
PPO	.2	.7	.1
FFS	.1	.3	.6

In 2000, the percentages for the three plans were *HMO*:30 %, *PPO*:25 %, and *FFS*:45 %. (a) What will be the percentages for the three plans in 2001? (b) What is the long run fraction choosing each of the three plans?

1.28. Bob eats lunch at the campus food court every week day. He either eats Chinese food, Quesadila, or Salad. His transition matrix is

	C	Q	S
C	.15	.6	.25
Q	.4	.1	.5
S	.1	.3	.6

He had Chinese food on Monday. (a) What are the probabilities for his three meal choices on Friday (four days later). (b) What are the long run frequencies for his three choices?

1.29. The liberal town of Ithaca has a "free bikes for the people program." You can pick up bikes at the library (L), the coffee shop (C), or the cooperative grocery store (G). The director of the program has determined that bikes move around according to the following Markov chain

$$
\begin{array}{c c c c}
 & L & C & G \\
L & .5 & .2 & .3 \\
C & .4 & .5 & .1 \\
G & .25 & .25 & .5
\end{array}
$$

On Sunday there are an equal number of bikes at each place. (a) What fraction of the bikes are at the three locations on Tuesday? (b) on the next Sunday? (c) In the long run what fraction are at the three locations?

1.30. A plant species has red, pink, or white flowers according to the genotypes RR, RW, and WW, respectively. If each of these genotypes is crossed with a pink (*RW*) plant, then the offspring fractions are

$$
\begin{array}{c c c c}
 & RR & RW & WW \\
RR & .5 & .5 & 0 \\
RW & .25 & .5 & .25 \\
WW & 0 & .5 & .5
\end{array}
$$

What is the long run fraction of plants of the three types?

1.31. The weather in a certain town is classified as rainy, cloudy, or sunny and changes according to the following transition probability is

$$
\begin{array}{c c c c}
 & R & C & S \\
R & 1/2 & 1/4 & 1/4 \\
C & 1/4 & 1/2 & 1/4 \\
S & 1/2 & 1/2 & 0
\end{array}
$$

In the long run what proportion of days in this town are rainy? cloudy? sunny?

1.32. A sociologist studying living patterns in a certain region determines that the pattern of movement between urban (U), suburban (S), and rural areas (R) is given by the following transition matrix.

$$
\begin{array}{c c c c}
 & U & S & R \\
U & .86 & .08 & .06 \\
S & .05 & .88 & .07 \\
R & .03 & .05 & .92
\end{array}
$$

In the long run what fraction of the population will live in the three areas?

1.33. In a large metropolitan area, commuters either drive alone (A), carpool (C), or take public transportation (T). A study showed that transportation changes according to the following matrix:

$$
\begin{array}{cccc}
 & \text{A} & \text{C} & \text{T} \\
\text{A} & .8 & .15 & .05 \\
\text{C} & .05 & .9 & .05 \\
\text{S} & .05 & .1 & .85 \\
\end{array}
$$

In the long run what fraction of commuters will use the three types of transportation?

1.34. (a) Three telephone companies A, B, and C compete for customers. Each year customers switch between companies according to the following transition probability

$$
\begin{array}{cccc}
 & \text{A} & \text{B} & \text{C} \\
\text{A} & .75 & .05 & .20 \\
\text{B} & .15 & .65 & .20 \\
\text{C} & .05 & .1 & .85 \\
\end{array}
$$

What is the limiting market share for each of these companies?

1.35. In a particular county voters declare themselves as members of the Republican, Democrat, or Green party. No voters change directly from the Republican to Green party or vice versa. Other transitions occur according to the following matrix:

$$
\begin{array}{cccc}
 & \text{R} & \text{D} & \text{G} \\
\text{R} & .85 & .15 & 0 \\
\text{D} & .05 & .85 & .10 \\
\text{G} & 0 & .05 & .95 \\
\end{array}
$$

In the long run what fraction of voters will belong to the three parties?

1.36. An auto insurance company classifies its customers into three categories: poor, satisfactory, and excellent. No one moves from poor to excellent or from excellent to poor in one year.

$$
\begin{array}{cccc}
 & \text{P} & \text{S} & \text{E} \\
\text{P} & .6 & .4 & 0 \\
\text{S} & .1 & .6 & .3 \\
\text{E} & 0 & .2 & .8 \\
\end{array}
$$

What is the limiting fraction of drivers in each of these categories?

1.37. A professor has two light bulbs in his garage. When both are burned out, they are replaced, and the next day starts with two working light bulbs. Suppose that when both are working, one of the two will go out with probability .02 (each has probability .01 and we ignore the possibility of losing two on the same day).

However, when only one is there, it will burn out with probability .05. (i) What is the long-run fraction of time that there is exactly one bulb working? (ii) What is the expected time between light bulb replacements?

1.38. An individual has three umbrellas, some at her office, and some at home. If she is leaving home in the morning (or leaving work at night) and it is raining, she will take an umbrella, if one is there. Otherwise, she gets wet. Assume that independent of the past, it rains on each trip with probability 0.2. To formulate a Markov chain, let X_n be the number of umbrellas at her current location. (a) Find the transition probability for this Markov chain. (b) Calculate the limiting fraction of time she gets wet.

1.39. Let X_n be the number of days since David last shaved, calculated at 7:30 a.m. when he is trying to decide if he wants to shave today. Suppose that X_n is a Markov chain with transition matrix

$$
\begin{array}{c|cccc}
 & 1 & 2 & 3 & 4 \\
\hline
1 & 1/2 & 1/2 & 0 & 0 \\
2 & 2/3 & 0 & 1/3 & 0 \\
3 & 3/4 & 0 & 0 & 1/4 \\
4 & 1 & 0 & 0 & 0 \\
\end{array}
$$

In words, if he last shaved k days ago, he will not shave with probability $1/(k+1)$. However, when he has not shaved for 4 days his mother orders him to shave, and he does so with probability 1. (a) What is the long-run fraction of time David shaves? (b) Does the stationary distribution for this chain satisfy the detailed balance condition?

1.40. *Reflecting random walk on the line.* Consider the points 1, 2, 3, 4 to be marked on a straight line. Let X_n be a Markov chain that moves to the right with probability 2/3 and to the left with probability 1/3, but subject this time to the rule that if X_n tries to go to the left from 1 or to the right from 4 it stays put. Find (a) the transition probability for the chain, and (b) the limiting amount of time the chain spends at each site.

1.41. At the end of a month, a large retail store classifies each of its customer's accounts according to current (0), 30–60 days overdue (1), 60–90 days overdue (2), more than 90 days (3). Their experience indicates that the accounts move from state to state according to a Markov chain with transition probability matrix:

$$
\begin{array}{c|cccc}
 & 0 & 1 & 2 & 3 \\
\hline
0 & .9 & .1 & 0 & 0 \\
1 & .8 & 0 & .2 & 0 \\
2 & .5 & 0 & 0 & .5 \\
3 & .1 & 0 & 0 & .9 \\
\end{array}
$$

In the long run what fraction of the accounts are in each category?

1.42. At the beginning of each day, a piece of equipment is inspected to determine its working condition, which is classified as state $1 =$ new, 2, 3, or $4 =$ broken. Suppose that a broken machine requires three days to fix. To incorporate this into the Markov chain we add states 5 and 6 and write the following transition matrix:

	1	2	3	4	5	6
1	.95	.05	0	0	0	0
2	0	.9	.1	0	0	0
3	0	0	.875	.125	0	0
4	0	0	0	0	1	0
5	0	0	0	0	0	1
6	1	0	0	0	0	0

(a) Find the fraction of time that the machine is working. (b) Suppose now that we have the option of performing preventative maintenance when the machine is in state 3, and that this maintenance takes one day and returns the machine to state 1. This changes the transition probability to

	1	2	3
1	.95	.05	0
2	0	.9	.1
3	1	0	0

Find the fraction of time the machine is working under this new policy.

1.43. *Landscape dynamics.* To make a crude model of a forest we might introduce states $0 =$ grass, $1 =$ bushes, $2 =$ small trees, $3 =$ large trees, and write down a transition matrix like the following:

	0	1	2	3
0	1/2	1/2	0	0
1	1/24	7/8	1/12	0
2	1/36	0	8/9	1/12
3	1/8	0	0	7/8

The idea behind this matrix is that if left undisturbed a grassy area will see bushes grow, then small trees, which of course grow into large trees. However, disturbances such as tree falls or fires can reset the system to state 0. Find the limiting fraction of land in each of the states.

Exit Distributions and Times

1.44. The Markov chain associated with a manufacturing process may be described as follows: A part to be manufactured will begin the process by entering step 1. After step 1, 20 % of the parts must be reworked, i.e., returned to step 1, 10 % of the parts are thrown away, and 70 % proceed to step 2. After step 2, 5 % of the parts must be returned to the step 1, 10 % to step 2, 5 % are scrapped, and 80 % emerge to be sold for a profit. (a) Formulate a four-state Markov chain with states 1, 2, 3, and 4 where 3 = a part that was scrapped and 4 = a part that was sold for a profit. (b) Compute the probability a part is scrapped in the production process.

1.45. A bank classifies loans as paid in full (F), in good standing (G), in arrears (A), or as a bad debt (B). Loans move between the categories according to the following transition probability:

$$
\begin{array}{c|cccc}
 & F & G & A & B \\
\hline
F & 1 & 0 & 0 & 0 \\
G & .1 & .8 & .1 & 0 \\
A & .1 & .4 & .4 & .1 \\
B & 0 & 0 & 0 & 1 \\
\end{array}
$$

What fraction of loans in good standing are eventually paid in full? What is the answer for those in arrears?

1.46. A warehouse has a capacity to hold four items. If the warehouse is neither full nor empty, the number of items in the warehouse changes whenever a new item is produced or an item is sold. Suppose that (no matter when we look) the probability that the next event is "a new item is produced" is 2/3 and that the new event is a "sale" is 1/3. If there is currently one item in the warehouse, what is the probability that the warehouse will become full before it becomes empty.

1.47. The Duke football team can Pass, Run, throw an Interception, or Fumble. Suppose the sequence of outcomes is Markov chain with the following transition matrix.

$$
\begin{array}{c|cccc}
 & P & R & I & F \\
\hline
P & 0.7 & 0.2 & 0.1 & 0 \\
R & 0.35 & 0.6 & 0 & 0.05 \\
I & 0 & 0 & 1 & 0 \\
F & 0 & 0 & 0 & 1 \\
\end{array}
$$

The first play is a pass. (a) What is the expected number of plays until a fumble or interception? (b) What is the probability the sequence of plays ends in an interception.

1.48. Six children (Dick, Helen, Joni, Mark, Sam, and Tony) play catch. If Dick has the ball, he is equally likely to throw it to Helen, Mark, Sam, and Tony. If Helen has the ball, she is equally likely to throw it to Dick, Joni, Sam, and Tony. If Sam has the ball, he is equally likely to throw it to Dick, Helen, Mark, and Tony. If either Joni or Tony gets the ball, they keep throwing it to each other. If Mark gets the ball, he runs away with it. (a) Find the transition probability and classify the states of the chain. (b) Suppose Dick has the ball at the beginning of the game. What is the probability Mark will end up with it?

1.49. To find the waiting time for *HHH* we let the state of our Markov chains be the number of consecutive heads we have at the moment. The transition probability is

	0	1	2	3
0	0.5	0.5	0	0
1	0.5	0	0.5	0
2	0.5	0	0	0.5
3	0	0	0	1

Find $E_0 T_3$.

(b) (10 points) Consider now the chain where the state gives the outcomes of the last three tosses. It has state space $\{HHH, HHT, \ldots TTT\}$. Use the fact that $E_{HHH} T_{HHH} = 8$ to find the answer to part (a).

1.50. To find the waiting time for *HTH* we let the state of our Markov chains be the part of the pattern we have so far. The transition probability is

	0	H	HT	HTH
0	0.5	0.5	0	0
H	0	0.5	0.5	0
HT	0.5	0	0	0.5
HTH	0	0	0	1

(a) Find $E_0 T_{HTH}$. (b) use the reasoning for part (b) of the previous exercise to conclude $E_0 T_{HHT} = 8$ and $E_0 T_{HTT} = 8/$

1.51. *Sucker bet.* Consider the following gambling game. Player 1 picks a three coin pattern (for example, *HTH*) and player 2 picks another (say *THH*). A coin is flipped repeatedly and outcomes are recorded until one of the two patterns appears. Somewhat surprisingly player 2 has a considerable advantage in this game. No matter what player 1 picks, player 2 can win with probability $\geq 2/3$. Suppose without loss of generality that player 1 picks a pattern that begins with H:

case	Player 1	Player 2	Prob. 2 wins
1	HHH	THH	7/8
2	HHT	THH	3/4
3	HTH	HHT	2/3
4	HTT	HHT	2/3

Verify the results in the table. You can do this by solving six equations in six unknowns but this is not the easiest way.

1.52. At the New York State Fair in Syracuse, Larry encounters a carnival game where for one dollar he may buy a single coupon allowing him to play a guessing game. On each play, Larry has an even chance of winning or losing a coupon. When he runs out of coupons he loses the game. However, if he can collect three coupons, he wins a surprise. (a) What is the probability Larry will win the surprise? (b) What is the expected number of plays he needs to win or lose the game.

1.53. The Megasoft company gives each of its employees the title of programmer (P) or project manager (M). In any given year 70 % of programmers remain in that position 20 % are promoted to project manager and 10 % are fired (state X). 95 % of project managers remain in that position while 5 % are fired. How long on the average does a programmer work before they are fired?

1.54. At a nationwide travel agency, newly hired employees are classified as beginners (*B*). Every six months the performance of each agent is reviewed. Past records indicate that transitions through the ranks to intermediate (I) and qualified (Q) are according to the following Markov chain, where F indicates workers that were fired:

$$
\begin{array}{c|cccc}
 & \mathbf{B} & \mathbf{I} & \mathbf{Q} & \mathbf{F} \\
\hline
\mathbf{B} & .45 & .4 & 0 & .15 \\
\mathbf{I} & 0 & .6 & .3 & .1 \\
\mathbf{Q} & 0 & 0 & 1 & 0 \\
\mathbf{F} & 0 & 0 & 0 & 1 \\
\end{array}
$$

(a) What fraction eventually become qualified? (b) What is the expected time until a beginner is fired or becomes qualified?

1.55. At a manufacturing plant, employees are classified as trainee (R), technician (T), or supervisor (S). Writing Q for an employee who quits we model their progress through the ranks as a Markov chain with transition probability

$$
\begin{array}{c|cccc}
 & \mathbf{R} & \mathbf{T} & \mathbf{S} & \mathbf{Q} \\
\hline
\mathbf{R} & .2 & .6 & 0 & .2 \\
\mathbf{T} & 0 & .55 & .15 & .3 \\
\mathbf{S} & 0 & 0 & 1 & 0 \\
\mathbf{Q} & 0 & 0 & 0 & 1 \\
\end{array}
$$

(a) What fraction of recruits eventually make supervisor? (b) What is the expected time until a trainee audits or becomes supervisor?

1.56. Customers shift between variable rate loans (V), thirty year fixed-rate loans (30), fifteen year fixed-rate loans (15), or enter the states paid in full (P), or foreclosed according to the following transition matrix:

$$
\begin{array}{c|ccccc}
 & \textbf{V} & \textbf{30} & \textbf{15} & \textbf{P} & \textbf{F} \\
\hline
\textbf{V} & .55 & .35 & 0 & .05 & .05 \\
\textbf{30} & .15 & .54 & .25 & .05 & .01 \\
\textbf{15} & .20 & 0 & .75 & .04 & .01 \\
\textbf{P} & 0 & 0 & 0 & 1 & 0 \\
\textbf{F} & 0 & 0 & 0 & 0 & 1 \\
\end{array}
$$

(a) For each of the three loan types find (a) the expected time until paid or foreclosed. (b) the probability the loan is paid.

1.57. 3. Two barbers and two chairs. Consider the following chain

$$
\begin{array}{c|ccccc}
 & \textbf{0} & \textbf{1} & \textbf{2} & \textbf{3} & \textbf{4} \\
\hline
\textbf{0} & 0 & 1 & 0 & 0 & 0 \\
\textbf{1} & 0.6 & 0 & 0.4 & 0 & 0 \\
\textbf{2} & 0 & 0.75 & 0 & 0.25 & 0 \\
\textbf{3} & 0 & 0 & 0.75 & 0 & 0.25 \\
\textbf{4} & 0 & 0 & 0 & 0.75 & 0.25 \\
\end{array}
$$

(a) Find the stationary distribution. (b) Compute $P_x(V_0 < V_4)$ for $x = 1, 2, 3$. (c) Let $\tau = \min\{V_0, V_4\}$. Find $E_x \tau$ for $x = 1, 2, 3$.

More Theoretical Exercises

1.58. Consider a general chain with state space $S = \{1, 2\}$ and write the transition probability as

$$
\begin{array}{c|cc}
 & \textbf{1} & \textbf{2} \\
\hline
\textbf{1} & 1-a & a \\
\textbf{2} & b & 1-b \\
\end{array}
$$

Use the Markov property to show that

$$
P(X_{n+1} = 1) - \frac{b}{a+b} = (1-a-b)\left\{ P(X_n = 1) - \frac{b}{a+b} \right\}
$$

and then conclude

$$P(X_n = 1) = \frac{b}{a+b} + (1-a-b)^n \left\{ P(X_0 = 1) - \frac{b}{a+b} \right\}$$

This shows that if $0 < a+b < 2$, then $P(X_n = 1)$ converges exponentially fast to its limiting value $b/(a+b)$.

1.59. *Bernoulli–Laplace model of diffusion.* Consider two urns each of which contains m balls; b of these $2m$ balls are black, and the remaining $2m-b$ are white. We say that the system is in state i if the first urn contains i black balls and $m-i$ white balls while the second contains $b-i$ black balls and $m-b+i$ white balls. Each trial consists of choosing a ball at random from each urn and exchanging the two. Let X_n be the state of the system after n exchanges have been made. X_n is a Markov chain. (a) Compute its transition probability. (b) Verify that the stationary distribution is given by

$$\pi(i) = \binom{b}{i}\binom{2m-b}{m-i} \bigg/ \binom{2m}{m}$$

(c) Can you give a simple intuitive explanation why the formula in (b) gives the right answer?

1.60. *Library chain.* On each request the ith of n possible books is the one chosen with probability p_i. To make it quicker to find the book the next time, the librarian moves the book to the left end of the shelf. Define the state at any time to be the sequence of books we see as we examine the shelf from left to right. Since all the books are distinct this list is a permutation of the set $\{1, 2, \ldots n\}$, i.e., each number is listed exactly once. Show that

$$\pi(i_1, \ldots, i_n) = p_{i_1} \cdot \frac{p_{i_2}}{1 - p_{i_1}} \cdot \frac{p_{i_3}}{1 - p_{i_1} - p_{i_2}} \cdots \frac{p_{i_n}}{1 - p_{i_1} - \cdots p_{i_{n-1}}}$$

is a stationary distribution.

1.61. *Random walk on a clock.* Consider the numbers $1, 2, \ldots 12$ written around a ring as they usually are on a clock. Consider a Markov chain that at any point jumps with equal probability to the two adjacent numbers. (a) What is the expected number of steps that X_n will take to return to its starting position? (b) What is the probability X_n will visit all the other states before returning to its starting position?

1.62. *King's random walk.* This and the next example continue Example 1.30. A king can move one squares horizontally, vertically, or diagonally. Let X_n be the sequence of squares that results if we pick one of king's legal moves at random. Find (a) the stationary distribution and (b) the expected number of moves to return to corner $(1,1)$ when we start there.

1.63. *Queen's random walk.* A queen can move any number of squares horizontally, vertically, or diagonally. Let X_n be the sequence of squares that results if we pick one of queen's legal moves at random. Find (a) the stationary distribution and (b) the expected number of moves to return to corner $(1,1)$ when we start there.

1.64. *Wright–Fisher model.* Consider the chain described in Example 1.7.

$$p(x, y) = \binom{N}{y}(\rho_x)^y(1 - \rho_x)^{N-y}$$

where $\rho_x = (1 - u)x/N + v(N - x)/N$. (a) Show that if $u, v > 0$, then $\lim_{n\to\infty} p^n(x, y) = \pi(y)$, where π is the unique stationary distribution. There is no known formula for $\pi(y)$, but you can (b) compute the mean $v = \sum_y y\pi(y) = \lim_{n\to\infty} E_x X_n$.

1.65. *Ehrenfest chain.* Consider the Ehrenfest chain, Example 1.2, with transition probability $p(i, i + 1) = (N - i)/N$, and $p(i, i - 1) = i/N$ for $0 \le i \le N$. Let $\mu_n = E_x X_n$. (a) Show that $\mu_{n+1} = 1 + (1 - 2/N)\mu_n$. (b) Use this and induction to conclude that

$$\mu_n = \frac{N}{2} + \left(1 - \frac{2}{N}\right)^n (x - N/2)$$

From this we see that the mean μ_n converges exponentially rapidly to the equilibrium value of $N/2$ with the error at time n being $(1 - 2/N)^n(x - N/2)$.

1.66. *Brother–sister mating.* In this genetics scheme two individuals (one male and one female) are retained from each generation and are mated to give the next. If the individuals involved are diploid and we are interested in a trait with two alleles, A and a, then each individual has three possible states AA, Aa, aa or more succinctly 2, 1, 0. If we keep track of the sexes of the two individuals the chain has nine states, but if we ignore the sex there are just six: 22, 21, 20, 11, 10, and 00. (a) Assuming that reproduction corresponds to picking one letter at random from each parent, compute the transition probability. (b) 22 and 00 are absorbing states for the chain. Show that the probability of absorption in 22 is equal to the fraction of A's in the state. (c) Let $T = \min\{n \ge 0 : X_n = 22 \text{ or } 00\}$ be the absorption time. Find $E_x T$ for all states x.

1.67. Roll a fair die repeatedly and let $Y_1, Y_2, \ldots$ be the resulting numbers. Let $X_n = |\{Y_1, Y_2, \ldots, Y_n\}|$ be the number of values we have seen in the first n rolls for $n \ge 1$ and set $X_0 = 0$. X_n is a Markov chain. (a) Find its transition probability. (b) Let $T = \min\{n : X_n = 6\}$ be the number of trials we need to see all 6 numbers at least once. Find ET.

1.68. *Coupon collector's problem.* We are interested now in the time it takes to collect a set of N baseball cards. Let T_k be the number of cards we have to buy before we have k that are distinct. Clearly, $T_1 = 1$. A little more thought reveals that if each time we get a card chosen at random from all N possibilities, then for $k \ge 1$,

$T_{k+1} - T_k$ has a geometric distribution with success probability $(N-k)/N$. Use this to show that the mean time to collect a set of N baseball cards is $\approx N \log N$, while the variance is $\approx N^2 \sum_{k=1}^{\infty} 1/k^2$.

1.69. *Algorithmic efficiency.* The simplex method minimizes linear functions by moving between extreme points of a polyhedral region so that each transition decreases the objective function. Suppose there are n extreme points and they are numbered in increasing order of their values. Consider the Markov chain in which $p(1,1) = 1$ and $p(i,j) = 1/i - 1$ for $j < i$. In words, when we leave j we are equally likely to go to any of the extreme points with better value. (a) Use (1.27) to show that for $i > 1$

$$E_i T_1 = 1 + 1/2 + \cdots + 1/(i-1)$$

(b) Let $I_j = 1$ if the chain visits j on the way from n to 1. Show that for $j < n$

$$P(I_j = 1 | I_{j+1}, \dots I_n) = 1/j$$

to get another proof of the result and conclude that $I_1, \dots I_{n-1}$ are independent.

Infinite State Space

1.70. *General birth and death chains.* The state space is $\{0, 1, 2, \dots\}$ and the transition probability has

$$
\begin{aligned}
p(x, x+1) &= p_x \\
p(x, x-1) &= q_x \qquad \text{for } x > 0 \\
p(x, x) &= r_x \qquad \text{for } x \geq 0
\end{aligned}
$$

while the other $p(x, y) = 0$. Let $V_y = \min\{n \geq 0 : X_n = y\}$ be the time of the first visit to y and let $h_N(x) = P_x(V_N < V_0)$. By considering what happens on the first step, we can write

$$h_N(x) = p_x h_N(x+1) + r_x h_N(x) + q_x h_N(x-1)$$

Set $h_N(1) = c_N$ and solve this equation to conclude that 0 is recurrent if and only if $\sum_{y=1}^{\infty} \prod_{x=1}^{y-1} q_x/p_x = \infty$ where by convention $\prod_{x=1}^{0} = 1$.

1.71. To see what the conditions in the last problem say we will now consider some concrete examples. Let $p_x = 1/2$, $q_x = e^{-cx^{-\alpha}}/2$, $r_x = 1/2 - q_x$ for $x \geq 1$ and $p_0 = 1$. For large x, $q_x \approx (1 - cx^{-\alpha})/2$, but the exponential formulation keeps the probabilities nonnegative and makes the problem easier to solve. Show that the chain is recurrent if $\alpha > 1$ or if $\alpha = 1$ and $c \leq 1$ but is transient otherwise.

1.72. Consider the Markov chain with state space $\{0, 1, 2, \ldots\}$ and transition probability

$$p(m, m+1) = \frac{1}{2}\left(1 - \frac{1}{m+2}\right) \qquad \text{for } m \geq 0$$

$$p(m, m-1) = \frac{1}{2}\left(1 + \frac{1}{m+2}\right) \qquad \text{for } m \geq 1$$

and $p(0, 0) = 1 - p(0, 1) = 3/4$. Find the stationary distribution π.

1.73. Consider the Markov chain with state space $\{1, 2, \ldots\}$ and transition probability

$$p(m, m+1) = m/(2m+2) \qquad \text{for } m \geq 1$$
$$p(m, m-1) = 1/2 \qquad \text{for } m \geq 2$$
$$p(m, m) = 1/(2m+2) \qquad \text{for } m \geq 2$$

and $p(1, 1) = 1 - p(1, 2) = 3/4$. Show that there is no stationary distribution.

1.74. Consider the aging chain on $\{0, 1, 2, \ldots\}$ in which for any $n \geq 0$ the individual gets one day older from n to $n+1$ with probability p_n but dies and returns to age 0 with probability $1 - p_n$. Find conditions that guarantee that (a) 0 is recurrent, (b) positive recurrent. (c) Find the stationary distribution.

1.75. The opposite of the aging chain is the renewal chain with state space $\{0, 1, 2, \ldots\}$ in which $p(i, i-1) = 1$ when $i > 0$. The only nontrivial part of the transition probability is $p(0, i) = p_i$. Show that this chain is always recurrent but is positive recurrent if and only if $\sum_n np_n < \infty$.

1.76. Consider a branching process as defined in 1.55, in which each family has exactly three children, but invert Galton and Watson's original motivation and ignore male children. In this model a mother will have an average of 1.5 daughters. Compute the probability that a given woman's descendents will die out.

1.77. Consider a branching process as defined in 1.55, in which each family has a number of children that follows a shifted geometric distribution: $p_k = p(1-p)^k$ for $k \geq 0$, which counts the number of failures before the first success when success has probability p. Compute the probability that starting from one individual the chain will be absorbed at 0.

Chapter 2
Poisson Processes

2.1 Exponential Distribution

To prepare for our discussion of the Poisson process, we need to recall the definition and some of the basic properties of the exponential distribution. A random variable T is said to have **an exponential distribution with rate** λ, or $T = \text{exponential}(\lambda)$, if

$$P(T \le t) = 1 - e^{-\lambda t} \quad \text{for all } t \ge 0 \tag{2.1}$$

Here we have described the distribution by giving the **distribution function** $F(t) = P(T \le t)$. We can also write the definition in terms of the **density function** $f_T(t)$ which is the derivative of the distribution function.

$$f_T(t) = \begin{cases} \lambda e^{-\lambda t} & \text{for } t \ge 0 \\ 0 & \text{for } t < 0 \end{cases} \tag{2.2}$$

Integrating by parts with $f(t) = t$ and $g'(t) = \lambda e^{-\lambda t}$,

$$ET = \int_0^\infty t \cdot \lambda e^{-\lambda t}\, dt$$

$$= -te^{-\lambda t}\big|_0^\infty + \int_0^\infty e^{-\lambda t}\, dt = 1/\lambda \tag{2.3}$$

Integrating by parts with $f(t) = t^2$ and $g'(t) = \lambda e^{-\lambda t}$, we see that

$$ET^2 = \int_0^\infty t^2 \cdot \lambda e^{-\lambda t}\, dt$$

$$= -t^2 e^{-\lambda t}\big|_0^\infty + \int_0^\infty 2t e^{-\lambda t}\, dt = 2/\lambda^2 \tag{2.4}$$

© Springer International Publishing Switzerland 2016
R. Durrett, *Essentials of Stochastic Processes*, Springer Texts in Statistics,
DOI 10.1007/978-3-319-45614-0_2

by the formula for ET. So the variance

$$\text{var}\,(T) = ET^2 - (ET)^2 = 1/\lambda^2 \tag{2.5}$$

While calculus is required to know the exact values of the mean and variance, it is easy to see how they depend on λ. Let $T = $ exponential(λ), i.e., have an exponential distribution with rate λ, and let $S = $ exponential(1). To see that S/λ has the same distribution as T, we use (2.1) to conclude

$$P(S/\lambda \le t) = P(S \le \lambda t) = 1 - e^{-\lambda t} = P(T \le t)$$

Recalling that if c is any number then $E(cX) = cEX$ and $\text{var}\,(cX) = c^2 \,\text{var}\,(X)$, we see that

$$ET = ES/\lambda \qquad \text{var}\,(T) = \text{var}\,(S)/\lambda^2$$

Lack of Memory Property It is traditional to formulate this property in terms of waiting for an unreliable bus driver. In words, "if we've been waiting for t units of time then the probability we must wait s more units of time is the same as if we haven't waited at all." In symbols

$$P(T > t + s | T > t) = P(T > s) \tag{2.6}$$

To prove this we recall that if $B \subset A$, then $P(B|A) = P(B)/P(A)$, so

$$P(T > t + s | T > t) = \frac{P(T > t + s)}{P(T > t)} = \frac{e^{-\lambda(t+s)}}{e^{-\lambda t}} = e^{-\lambda s} = P(T > s)$$

where in the third step we have used the fact $e^{a+b} = e^a e^b$.

Exponential Races Let $S = $ exponential(λ) and $T = $ exponential(μ) be independent. In order for the minimum of S and T to be larger than t, each of S and T must be larger than t. Using this and independence we have

$$P(\min(S,T) > t) = P(S > t, T > t) = P(S > t)P(T > t)$$

$$= e^{-\lambda t} e^{-\mu t} = e^{-(\lambda+\mu)t} \tag{2.7}$$

That is, $\min(S,T)$ has an exponential distribution with rate $\lambda + \mu$.

We will now consider: "Who finishes first?" Breaking things down according to the value of S and then using independence with our formulas (2.1) and (2.2) for the distribution and density functions, to conclude

$$P(S < T) = \int_0^\infty f_S(s) P(T > s)\, ds$$

$$= \int_0^\infty \lambda e^{-\lambda s} e^{-\mu s} \, ds$$

$$= \frac{\lambda}{\lambda + \mu} \int_0^\infty (\lambda + \mu) e^{-(\lambda+\mu)s} \, ds = \frac{\lambda}{\lambda + \mu} \tag{2.8}$$

where on the last line we have used the fact that $(\lambda+\mu)e^{-(\lambda+\mu)s}$ is a density function and hence must integrate to 1.

Example 2.1. Anne and Betty enter a beauty parlor simultaneously, Anne to get a manicure and Betty to get a haircut. Suppose the time for a manicure (haircut) is exponentially distributed with mean 20 (30) minutes. (a) What is the probability Anne gets done first? (b) What is the expected amount of time until Anne and Betty are both done?

(a) The rates are 1/20 and 1/30 per hour so Anne finishes first with probability

$$\frac{1/20}{1/20 + 1/30} = \frac{30}{30 + 20} = \frac{3}{5}$$

(b) The total service rate is $1/30+1/20 = 5/60$, so the time until the first customer completes service is exponential with mean 12 minutes. With probability 3/5, Anne is done first. When this happens the lack of memory property of the exponential implies that it will take an average of 30 minutes for Betty to complete her haircut. With probability 2/5's Betty is done first and Anne will take an average of 20 more minutes. Combining we see that the total waiting time is

$$12 + (3/5) \cdot 30 + (2/5) \cdot 20 = 12 + 18 + 8 = 38$$

Races, II. n Random Variables The last two calculations extend easily to a sequence of independent random variables $T_i = $ exponential(λ_i), $1 \le i \le n$.

Theorem 2.1. *Let $V = \min(T_1, \ldots, T_n)$ and I be the (random) index of the T_i that is smallest.*

$$P(V > t) = \exp(-(\lambda_1 + \cdots + \lambda_n)t)$$

$$P(I = i) = \frac{\lambda_i}{\lambda_1 + \cdots + \lambda_n}$$

I and $V = \min\{T_1, \ldots T_n\}$ are independent.

Proof. Arguing as in the case of two random variables:

$$P(\min(T_1, \ldots, T_n) > t) = P(T_1 > t, \ldots T_n > t)$$

$$= \prod_{i=1}^n P(T_i > t) = \prod_{i=1}^n e^{-\lambda_i t} = e^{-(\lambda_1 + \cdots + \lambda_n)t}$$

That is, the minimum, $\min(T_1, \ldots, T_n)$, of several independent exponentials has an exponential distribution with rate equal to the sum of the rates $\lambda_1 + \cdots \lambda_n$.

To prove the second result let $S = T_i$ and U be the minimum of $T_j, j \neq i$. (2.1) implies that U is exponential with parameter

$$\mu = (\lambda_1 + \cdots + \lambda_n) - \lambda_i$$

so using the result for two random variables

$$P(T_i = \min(T_1, \ldots, T_n)) = P(S < U) = \frac{\lambda_i}{\lambda_i + \mu} = \frac{\lambda_i}{\lambda_1 + \cdots + \lambda_n}$$

To prove the independence let I be the (random) index of the T_i that is smallest. Let $f_{i,V}(t)$ be the density function for V on the set $I = i$. In order for i to be first at time t, $T_i = t$ and the other $T_j > t$ so

$$f_{i,V}(t) = \lambda_i e^{-\lambda_i t} \cdot \prod_{j \neq i} e^{-\lambda_j t}$$

$$= \frac{\lambda_i}{\lambda_1 + \cdots + \lambda_n} \cdot (\lambda_1 + \cdots + \lambda_n) e^{-(\lambda_1 + \cdots + \lambda_n)t}$$

$$= P(I = i) \cdot f_V(t)$$

since V has an exponential$(\lambda_1 + \cdots + \lambda_n)$ distribution.   □

Example 2.2. A submarine has three navigational devices but can remain at sea if at least two are working. Suppose that the failure times are exponential with means 1 year, 1.5 years, and 3 years. (a) What is the average length of time the boat can remain at sea? (b) Call the parts A, B, and C. Find the probabilities for the six orders in which the failures can occur.

(a) The rates for the three exponentials are $1, 2/3$, and $1/3$, per year. Thus the time to the first failure is exponential with rate $2 = 1 + 2/3 + 1/3$, so the mean time to first failure is $1/2$. $1/2$ of the time part 1 is the first to fail. In this case the time to the next failure has rate $2/3 + 1/3 = 1$ so the mean is 1. Part 2 is the first to fail with probability $2/6$. In this case the time to the next failure has rate $1 + 1/3 = 4/3$ or mean $3/4$. Part 3 is the first to fail with probability $1/6$. In this case the time to the next failure has rate $1 + 2/3 = 5/3$ and mean $3/5$. Adding things up the mean time until the second failure is

$$1/2 + (1/2) \cdot 1 + (1/3) \cdot (3/4) + (1/6) \cdot (3/5)$$

$$= .5 + .5 + .25 + .10 = 1.35 \text{ years}$$

For (b) the arithmetic is easier if we think of the rates as 3, 2, and 1.

$$ABC \ (1/2)(2/3) = 1/3 = 20/60$$
$$ACB \ (1/2)(1/3) = 1/6 = 10/60$$
$$BAC \ (1/3)(3/4) = 1/4 = 15/60$$
$$BCA \ (1/3)(1/4) = 1/12 = 5/60$$
$$CAB \ (1/6)(3/5) = 1/10 = 6/60$$
$$CBA \ (1/6)(2/5) = 2/30 = 4/60$$

Our final fact in this section concerns sums of exponentials.

Theorem 2.2. *Let $\tau_1, \tau_2, \ldots$ be independent exponential(λ). The sum $T_n = \tau_1 + \cdots + \tau_n$ has a gamma(n, λ) distribution. That is, the density function of T_n is given by*

$$f_{T_n}(t) = \lambda e^{-\lambda t} \cdot \frac{(\lambda t)^{n-1}}{(n-1)!} \quad \text{for } t \geq 0 \tag{2.9}$$

and 0 otherwise.

Proof. The proof is by induction on n. When $n = 1$, T_1 has an exponential(λ) distribution. Recalling that the 0th power of any positive number is 1, and by convention we set $0! = 1$, the formula reduces to

$$f_{T_1}(t) = \lambda e^{-\lambda t}$$

and we have shown that our formula is correct for $n = 1$.

To do the induction step, suppose that the formula is true for n. The sum $T_{n+1} = T_n + \tau_{n+1}$, so breaking things down according to the value of T_n, and using the independence of T_n and τ_{n+1}, we have

$$f_{T_{n+1}}(t) = \int_0^t f_{T_n}(s) f_{\tau_{n+1}}(t - s) \, ds$$

Plugging the formula from (2.9) in for the first term and the exponential density in for the second and using the fact that $e^a e^b = e^{a+b}$ with $a = -\lambda s$ and $b = -\lambda(t - s)$ gives

$$\int_0^t \lambda e^{-\lambda s} \frac{(\lambda s)^{n-1}}{(n-1)!} \cdot \lambda e^{-\lambda(t-s)} \, ds = e^{-\lambda t} \lambda^n \int_0^t \frac{s^{n-1}}{(n-1)!} \, ds$$

$$= \lambda e^{-\lambda t} \frac{\lambda^n t^n}{n!}$$

which completes the proof. □

2.2 Defining the Poisson Process

In order to prepare for the definition of the Poisson process, we introduce the
Poisson distribution and derive some of its properties.

Definition. We say that X has a **Poisson distribution** with mean λ, or $X =$
Poisson(λ), for short, if

$$P(X = n) = e^{-\lambda} \frac{\lambda^n}{n!} \quad \text{for } n = 0, 1, 2, \ldots$$

The next result computes the moments of the Poisson.

Theorem 2.3. *For any $k \geq 1$*

$$EX(X - 1) \cdots (X - k + 1) = \lambda^k \tag{2.10}$$

and hence $var(X) = \lambda$

Proof. $X(X - 1) \cdots (X - k + 1) = 0$ if $X \leq k - 1$ so

$$EX(X - 1) \cdots (X - k + 1) = \sum_{j=k}^{\infty} e^{-\lambda} \frac{\lambda^j}{j!} j(j - 1) \cdots (j - k + 1)$$

$$= \lambda^k \sum_{j=k}^{\infty} e^{-\lambda} \frac{\lambda^{j-k}}{(j - k)!} = \lambda^k$$

since the sum gives the total mass of the Poisson distribution. Using $var(X) =$
$E(X(X - 1)) + EX - (EX)^2$ we conclude

$$var(X) = \lambda^2 + \lambda - (\lambda)^2 = \lambda$$

□

Theorem 2.4. *If X_i are independent Poisson(λ_i), then*

$$X_1 + \cdots + X_k = Poisson(\lambda_1 + \cdots + \lambda_n).$$

Proof. It suffices to prove the result for $k = 2$, for then the general result follows by
induction.

$$P(X_1 + X_2 = n) = \sum_{m=0}^{n} P(X_1 = m)P(X_2 = n - m)$$

$$= \sum_{m=0}^{n} e^{-\lambda_1} \frac{(\lambda_1)^m}{m!} \cdot e^{-\lambda_2} \frac{(\lambda_2)^{n-m}}{(n - m)!}$$

Knowing the answer we want, we can rewrite the last expression as

$$e^{-(\lambda_1+\lambda_2)}\frac{(\lambda_1+\lambda_2)^n}{n!} \cdot \sum_{m=0}^{n} \binom{n}{m} \left(\frac{\lambda_1}{\lambda_1+\lambda_2}\right)^m \left(\frac{\lambda_2}{\lambda_1+\lambda_2}\right)^{n-m}$$

The sum is 1, since it is the sum of all the probabilities for a binomial(n,p) distribution with $p = \lambda_1/(\lambda_1+\lambda_2)$. The term outside the sum is the desired Poisson probability, so have proved the desired result. $\square$

We are now ready to define the Poisson process. To do this think about people arriving to use an ATM, and let $N(s)$ be the number of arrivals in $[0,s]$.

Definition. $\{N(s), s \geq 0\}$ is a Poisson process, if (i) $N(0) = 0$,

(ii) $N(t+s) - N(s) = \text{Poisson}(\lambda t)$, and
(iii) $N(t)$ has **independent increments**, i.e., if $t_0 < t_1 < \ldots < t_n$ then

$$N(t_1) - N(t_0), \ldots N(t_n) - N(t_{n-1}) \quad \text{are independent.}$$

To motivate the definition, suppose that each of the $n \approx 7000$ undergraduate students on Duke campus flips a coin with probability λ/n of heads to decide if they will go to an ATM in the Bryan Center between 12:17 and 12:18. The probability that exactly k students will go during the one-minute time interval is given by the binomial$(n, \lambda/n)$ distribution

$$\frac{n(n-1)\cdots(n-k+1)}{k!} \left(\frac{\lambda}{n}\right)^k \left(1-\frac{\lambda}{n}\right)^{n-k} \tag{2.11}$$

Theorem 2.5. *If n is large, the binomial$(n, \lambda/n)$ distribution is approximately Poisson(λ).*

Proof. Exchanging the numerators of the first two fractions and breaking the last term into two, (2.11) becomes

$$\frac{\lambda^k}{k!} \cdot \frac{n(n-1)\cdots(n-k+1)}{n^k} \cdot \left(1-\frac{\lambda}{n}\right)^n \left(1-\frac{\lambda}{n}\right)^{-k} \tag{2.12}$$

Considering the four terms separately, we have

(i) $\lambda^k/k!$ does not depend on n.
(ii) There are k terms on the top and k terms on the bottom, so we can write this fraction as

$$\frac{n}{n} \cdot \frac{n-1}{n} \cdots \frac{n-k+1}{n}$$

For any j we have $(n-j)/n \to 1$ as $n \to \infty$, so the second term converges to 1 as $n \to \infty$.

(iii) Skipping to the last term in (2.12), $\lambda/n \to 0$, so $1 - \lambda/n \to 1$. The power $-k$ is fixed so

$$\left(1 - \frac{\lambda}{n}\right)^{-k} \to 1^{-k} = 1$$

(iv) We broke off the last piece to make it easier to invoke one of the famous facts of calculus:

$$(1 - \lambda/n)^n \to e^{-\lambda} \quad \text{as } n \to \infty.$$

If you haven't seen this before, recall that

$$\log(1-x) = -x + x^2/2 + \ldots$$

so we have $n \log(1 - \lambda/n) = -\lambda + \lambda^2/n + \ldots \to \lambda$ as $n \to \infty$.

Combining (i)–(iv), we see that (2.12) converges to

$$\frac{\lambda^k}{k!} \cdot 1 \cdot e^{-\lambda} \cdot 1$$

which is the Poisson distribution with mean λ. □

By extending the last argument we can also see why the number of individuals that arrive in two disjoint time intervals should be independent. Using the multinomial instead of the binomial, we see that the probability j people will go between 12:17 and 12:18 and k people will go between 12:18 and 12:20 is

$$\frac{n!}{j!k!(n-j-k)!} \left(\frac{\lambda}{n}\right)^j \left(\frac{2\lambda}{n}\right)^k \left(1 - \frac{3\lambda}{n}\right)^{n-(j+k)}$$

Rearranging gives

$$\frac{(\lambda)^j}{j!} \cdot \frac{(2\lambda)^k}{k!} \cdot \frac{n(n-1)\cdots(n-j-k+1)}{n^{j+k}} \cdot \left(1 - \frac{3\lambda}{n}\right)^{n-(j+k)}$$

Reasoning as before shows that when n is large, this is approximately

$$\frac{(\lambda)^j}{j!} \cdot \frac{(2\lambda)^k}{k!} \cdot 1 \cdot e^{-3\lambda}$$

Writing $e^{-3\lambda} = e^{-\lambda}e^{-2\lambda}$ and rearranging we can write the last expression as

$$e^{-\lambda}\frac{\lambda^j}{j!} \cdot e^{-2\lambda}\frac{(2\lambda)^k}{k!}$$

This shows that the number of arrivals in the two time intervals we chose are independent Poissons with means λ and 2λ.

The last proof can be easily generalized to show that if we divide the hour between 12:00 and 1:00 into any number of intervals, then the arrivals are independent Poissons with the right means. However, the argument gets very messy to write down.

2.2.1 Constructing the Poisson Process

Definition. Let $\tau_1, \tau_2, \ldots$ be independent exponential(λ) random variables. Let $T_n = \tau_1 + \cdots + \tau_n$ for $n \geq 1$, $T_0 = 0$, and define $N(s) = \max\{n : T_n \leq s\}$.

We think of the τ_n as times between arrivals of customers at the ATM, so $T_n = \tau_1 + \cdots + \tau_n$ is the arrival time of the nth customer, and $N(s)$ is the number of arrivals by time s. To check the last interpretation, consider the following example:

and note that $N(s) = 4$ when $T_4 \leq s < T_5$, that is, the 4th customer has arrived by time s but the 5th has not.

To show that this constructs the Poisson process we begin by checking (ii).

Lemma 2.6. $N(s)$ *has a Poisson distribution with mean* λs.

Proof. Now $N(s) = n$ if and only if $T_n \leq s < T_{n+1}$; i.e., the nth customer arrives before time s but the $(n+1)$th after s. Breaking things down according to the value of $T_n = t$ and noting that for $T_{n+1} > s$, we must have $\tau_{n+1} > s - t$, and τ_{n+1} is independent of T_n, it follows that

$$P(N(s) = n) = \int_0^s f_{T_n}(t) P(t_{n+1} > s - t)\, dt$$

Plugging in (2.9) now, the last expression is

$$= \int_0^s \lambda e^{-\lambda t} \frac{(\lambda t)^{n-1}}{(n-1)!} \cdot e^{-\lambda(s-t)}\, dt$$

$$= \frac{\lambda^n}{(n-1)!} e^{-\lambda s} \int_0^s t^{n-1}\, dt = e^{-\lambda s} \frac{(\lambda s)^n}{n!}$$

which proves the desired result.   □

The key to proving (iii) is the following Markov property:

Lemma 2.7. $N(t+s) - N(s)$, $t \geq 0$ is a rate λ Poisson process and independent of $N(r)$, $0 \leq r \leq s$.

Why is this true? Suppose for concreteness (and so that we can reuse the last picture) that by time s there have been four arrivals T_1, T_2, T_3, T_4 that occurred at times t_1, t_2, t_3, t_4. We know that the waiting time for the fifth arrival must have $\tau_5 > s - t_4$, but by the lack of memory property of the exponential distribution (2.6)

$$P(\tau_5 > s - t_4 + t | \tau_5 > s - t_4) = P(\tau_5 > t) = e^{-\lambda t}$$

This shows that the distribution of the first arrival after s is exponential(λ) and independent of T_1, T_2, T_3, T_4. It is clear that $\tau_6, \tau_7, \ldots$ are independent of T_1, T_2, T_3, T_4, and τ_5. This shows that the interarrival times after s are independent exponential(λ), and hence that $N(t+s) - N(s)$, $t \geq 0$ is a Poisson process. $\square$

From Lemma 2.7 we get easily the following:

Lemma 2.8. $N(t)$ has independent increments.

Why is this true? Lemma 2.7 implies that $N(t_n) - N(t_{n-1})$ is independent of $N(r)$, $r \leq t_{n-1}$ and hence of $N(t_{n-1}) - N(t_{n-2}), \ldots N(t_1) - N(t_0)$. The desired result now follows by induction. $\square$

2.2.2 More Realistic Models

Two of the weaknesses of the derivation above are

(i) All students are assumed to have exactly the same probability of going to the ATM.
(ii) The probability of going in a given time interval is a constant multiple of the length of the interval, so the arrival rate of customers is constant during the day, i.e., the same at 1 p.m. and at 1 a.m.

(i) is a very strong assumption but can be weakened by using a more general Poisson approximation result like the following:

Theorem 2.9. Let $X_{n,m}$, $1 \leq m \leq n$ be independent random variables with $P(X_m = 1) = p_m$ and $P(X_m = 0) = 1 - p_m$. Let

$$S_n = X_1 + \cdots + X_n, \quad \lambda_n = ES_n = p_1 + \cdots + p_n,$$

and $Z_n = Poisson(\lambda_n)$. Then for any set A

$$|P(S_n \in A) - P(Z_n \in A)| \leq \sum_{m=1}^{n} p_m^2$$

Why is this true? If X and Y are integer valued random variables, then for any set A

$$|P(X \in A) - P(Y \in A)| \leq \frac{1}{2} \sum_n |P(X = n) - P(Y = n)|$$

The right-hand side is called the **total variation distance** between the two distributions and is denoted $\|X - Y\|$. If $P(X = 1) = p$, $P(X = 0) = 1 - p$, and $Y = \text{Poisson}(p)$, then

$$\sum_n |P(X = n) - P(Y = n)| = |(1 - p) - e^{-p}| + |p - pe^{-p}| + 1 - (1 + p)e^{-p}$$

Since $1 \geq e^{-p} \geq 1 - p$ the right-hand side is

$$e^{-p} - 1 + p + p - pe^{-p} + 1 - e^{-p} - pe^{-p} = 2p(1 - e^{-p} \leq 2p^2$$

Let $Y_m = \text{Poisson}(p_m)$ be independent. At this point we have shown $\|X_i - Y_i\| \leq p_i^2$. With a little work one can show

$$\|(X_1 + \cdots + X_n) - (Y_1 + \cdots + Y_n)\|$$

$$\leq \|(X_1, \cdots, X_n) - (Y_1, \cdots, Y_n)\| \leq \sum_{m=1}^n \|X_m - Y_m\|$$

and the desired result follows.   □

Theorem 2.9 is useful because it gives a bound on the difference between the distribution of S_n and the Poisson distribution with mean $\lambda_n = ES_n$. To bound the bound it is useful to note that

$$\sum_{m=1}^n p_m^2 \leq \max_k p_k \left(\sum_{m=1}^n p_m \right)$$

so the approximation is good if $\max_k p_k$ is small. This is similar to the usual heuristic for the normal distribution: the sum is due to small contributions from a large number of variables. However, here small means that it is nonzero with small probability. When a contribution is made it is equal to 1.

The last results handle problem (i). To address the problem of varying arrival rates mentioned in (ii), we generalize the definition.

Nonhomogeneous Poisson Processes *We say that $\{N(s), s \geq 0\}$ is a Poisson process with rate $\lambda(r)$ if (i) $N(0) = 0$,*

(ii) $N(t)$ has independent increments, and
(iii) $N(t) - N(s)$ is Poisson with mean $\int_s^t \lambda(r)\, dr$.

In this case, the interarrival times are not exponential and they are not independent. To demonstrate the first claim, we note that

$$P(\tau_1 > t) = P(N(t) = 0) = e^{-\int_0^t \lambda(s)\,ds}$$

since $N(t)$ is Poisson with mean $\mu(t) = \int_0^t \lambda(s)\,ds$. Differentiating gives the density function

$$P(\tau_1 = t) = -\frac{d}{dt}P(t_1 > t) = \lambda(t)e^{-\int_0^t \lambda(s)\,ds} = \lambda(t)e^{-\mu(t)}$$

Generalizing the last computation shows that the joint distribution

$$f_{T_1,T_2}(u, v) = \lambda(u)e^{-\mu(u)} \cdot \lambda(v)e^{-(\mu(v)-\mu(u))}$$

Changing variables, $s = u, t = v - u$, the joint density

$$f_{\tau_1,\tau_2}(s, t) = \lambda(s)e^{-\mu(s)} \cdot \lambda(s+t)e^{-(\mu(s+t)-\mu(s))}$$

so τ_1 and τ_2 are not independent when $\lambda(s)$ is not constant.

We will see a concrete example of a nonhomogeneous Poisson process in Example 2.9.

2.3 Compound Poisson Processes

In this section we will embellish our Poisson process by associating an independent and identically distributed (i.i.d.) random variable Y_i with each arrival. By independent we mean that the Y_i are independent of each other and of the Poisson process of arrivals. To explain why we have chosen these assumptions, we begin with two examples for motivation.

Example 2.3. Consider the McDonald's restaurant on Route 13 in the southern part of Ithaca. By arguments in the last section, it is not unreasonable to assume that between 12:00 and 1:00 cars arrive according to a Poisson process with rate λ. Let Y_i be the number of people in the ith vehicle. There might be some correlation between the number of people in the car and the arrival time, e.g., more families come to eat there at night, but for a first approximation it seems reasonable to assume that the Y_i are i.i.d. and independent of the Poisson process of arrival times.

Example 2.4. Messages arrive at a computer to be transmitted across the Internet. If we imagine a large number of users writing emails on their laptops (or tablets or smart phones), then the arrival times of messages can be modeled by a Poisson process. If we let Y_i be the size of the ith message, then again it is reasonable to assume $Y_1, Y_2, \ldots$ are i.i.d. and independent of the Poisson process of arrival times.

Having introduced the Y_i's, it is natural to consider the sum of the Y_i's we have seen up to time t:

$$S(t) = Y_1 + \cdots + Y_{N(t)}$$

where we set $S(t) = 0$ if $N(t) = 0$. In Example 2.3, $S(t)$ gives the number of customers that have arrived up to time t. In Example 2.4, $S(t)$ represents the total number of bytes in all of the messages up to time t. In either case it is interesting to know the mean and variance of $S(t)$.

Theorem 2.10. *Let $Y_1, Y_2, \ldots$ be independent and identically distributed, let N be an independent nonnegative integer valued random variable, and let $S = Y_1 + \cdots + Y_N$ with $S = 0$ when $N = 0$.*

(i) If $E|Y_i|$, $EN < \infty$, then $ES = EN \cdot EY_i$.
(ii) If EY_i^2, $EN^2 < \infty$, then $var(S) = EN\, var(Y_i) + var(N)(EY_i)^2$.
(iii) If N is Poisson(λ), then $var(S) = \lambda EY_i^2$.

Why is this reasonable? The first of these is natural since if $N = n$ is nonrandom $ES = nEY_i$. (i) then results by setting $n = EN$. The formula in (ii) is more complicated but it clearly has two of the necessary properties:

If $N = n$ is nonrandom, $var(S) = n\, var(Y_i)$.
If $Y_i = c$ is nonrandom $var(S) = c^2\, var(N)$.

Combining these two observations, we see that $EN\, var(Y_i)$ is the contribution to the variance from the variability of the Y_i, while $var(N)(EY_i)^2$ is the contribution from the variability of N.

Proof. When $N = n$, $S = X_1 + \cdots + X_n$ has $ES = nEY_i$. Breaking things down according to the value of N,

$$ES = \sum_{n=0}^{\infty} E(S|N = n) \cdot P(N = n)$$

$$= \sum_{n=0}^{\infty} nEY_i \cdot P(N = n) = EN \cdot EY_i$$

For the second formula we note that when $N = n$, $S = X_1 + \cdots + X_n$ has $var(S) = n\, var(Y_i)$ and hence,

$$E(S^2|N = n) = n\, var(Y_i) + (nEY_i)^2$$

Computing as before we get

$$ES^2 = \sum_{n=0}^{\infty} E(S^2|N = n) \cdot P(N = n)$$

$$= \sum_{n=0}^{\infty} \{n \cdot \text{var}\,(Y_i) + n^2 (EY_i)^2\} \cdot P(N = n)$$

$$= (EN) \cdot \text{var}\,(Y_i) + EN^2 \cdot (EY_i)^2$$

To compute the variance now, we observe that

$$\text{var}\,(S) = ES^2 - (ES)^2$$

$$= (EN) \cdot \text{var}\,(Y_i) + EN^2 \cdot (EY_i)^2 - (EN \cdot EY_i)^2$$

$$= (EN) \cdot \text{var}\,(Y_i) + \text{var}\,(N) \cdot (EY_i)^2$$

where in the last step we have used $\text{var}\,(N) = EN^2 - (EN)^2$ to combine the second and third terms.

For part (iii), we note that in the special case of the Poisson, we have $EN = \lambda$ and $\text{var}\,(N) = \lambda$, so the result follows from $\text{var}\,(Y_i) + (EY_i)^2 = EY_i^2$. □

For a concrete example of the use of Theorem 2.10 consider

Example 2.5. Suppose that the number of customers at a liquor store in a day has a Poisson distribution with mean 81 and that each customer spends an average of \$8 with a standard deviation of \$6. It follows from (i) in Theorem 2.10 that the mean revenue for the day is $81 \cdot \$8 = \$648$. Using (iii), we see that the variance of the total revenue is

$$81 \cdot \{(\$6)^2 + (\$8)^2\} = \$8100$$

Taking square roots we see that the standard deviation of the revenue is \$90 compared with a mean of \$648.

2.4 Transformations

2.4.1 Thinning

In the previous section, we added up the Y_i's associated with the arrivals in our Poisson process to see how many customers, etc., we had accumulated by time t. In this section we will use the Y_i to split the Poisson process into several. Let $N_j(t)$ be the number of $i \le N(t)$ with $Y_i = j$. In Example 2.3, where Y_i is the number of people in the ith car, $N_j(t)$ will be the number of cars that have arrived by time t with exactly j people. The somewhat remarkable fact is

Theorem 2.11. $N_j(t)$ *are independent rate* $\lambda P(Y_i = j)$ *Poisson processes.*

Why is this remarkable? There are two "surprises" here: the resulting processes are Poisson and they are independent. To drive the point home consider a Poisson process with rate 10 per hour, and then flip coins to determine whether the arriving customers are male or female. One might think that seeing 40 men arrive in one hour would be indicative of a large volume of business and hence a larger than normal number of women, but Theorem 2.11 tells us that the number of men and the number of women that arrive per hour are independent.

Proof. To begin we suppose that $P(Y_i = 1) = p$ and $P(Y_i = 2) = 1 - p$, so there are only two Poisson processes to consider: $N_1(t)$ and $N_2(t)$. It should be clear that the independent increments property of the Poisson process implies that the pairs of increments

$$(N_1(t_i) - N_1(t_{i-1}), N_2(t_i) - N_2(t_{i-1})), \quad 1 \le i \le n$$

are independent of each other. Since $N_1(0) = N_2(0) = 0$ by definition, it only remains to check that the components $X_i = N_i(t + s) - N_i(s)$ are independent and have the right Poisson distributions. To do this, we note that if $X_1 = j$ and $X_2 = k$, then there must have been $j + k$ arrivals between s and $s + t$, j of which were assigned 1's and k of which were assigned 2's, so

$$P(X_1 = j, X_2 = k) = e^{-\lambda t} \frac{(\lambda t)^{j+k}}{(j+k)!} \cdot \frac{(j+k)!}{j!k!} p^j (1-p)^k$$

$$= e^{-\lambda pt} \frac{(\lambda pt)^j}{j!} e^{-\lambda(1-p)t} \frac{(\lambda(1-p)t)^k}{k!} \tag{2.13}$$

so $X_1 = \text{Poisson}(\lambda pt)$ and $X_2 = \text{Poisson}(\lambda(1-p)t)$. For the general case, we use the multinomial to conclude that if $p_j = P(Y_i = j)$ for $1 \le j \le m$ then

$$P(X_1 = k_1, \ldots X_m = k_m)$$

$$= e^{-\lambda t} \frac{(\lambda t)^{k_1 + \cdots k_m}}{(k_1 + \cdots k_m)!} \frac{(k_1 + \cdots k_m)!}{k_1! \cdots k_m!} p_1^{k_1} \cdots p_m^{k_m}$$

$$= \prod_{j=1}^{m} e^{-\lambda p_j t} \frac{(\lambda p_j)^{k_j}}{k_j!}$$

which proves the desired result. □

Example 2.6. Ellen catches fish at times of a Poisson process with rate 2 per hour. 40 % of the fish are salmon, while 60 % of the fish are trout. What is the probability she will catch exactly 1 salmon and 2 trout if she fishes for 2.5 hours? The total number of fish she catches in 2.5 hours is Poisson with mean 5, so the number of salmon and the number of trout are independent Poissons with means 2 and 3. Thus the probability of interest is

$$e^{-2} \frac{2^1}{1!} \cdot e^{-3} \frac{3^2}{2!}$$

Example 2.7. Two copy editors read a 300-page manuscript. The first found 100 typos, the second found 120, and their lists contain 80 errors in common. Suppose that the author's typos follow a Poisson process with some unknown rate λ per page, while the two copy editors catch errors with unknown probabilities of success p_1 and p_2. The goal in this problem is to find an estimate of λ. we want to estimate λ, p_1, p_2 and the number of undiscovered typos.

Let X_0 be the number of typos that neither found. Let X_1 and X_2 be the number of typos found only by 1 or only by 2, and let X_3 be the number of typos found by both. If we let $\mu = 300\lambda$, the X_i are independent Poisson with means

$$\mu(1 - p_1)(1 - p_2), \quad \mu p_1(1 - p_2), \quad \mu(1 - p_1)p_2, \quad \mu p_1 p_2$$

In our example $X_1 = 20$, $X_2 = 40$ and $X_3 = 80$. Since $EX_3/E(X_2 + X_3) = p_1$, solving gives $p_1 = 2/3$. $EX_3/E(X_1 + X_3) = p_2$ so $p_2 = 0.8$. Since $EX_0/EX_1 = (1 - p_1)/p_1 = EX_2/EX_3$ we guess that there are $20 \cdot 40/80 = 10$ typos remaining. Alternatively one can estimate $\mu = 80/p_1 p_2 = 150$, i.e., $\lambda = 1/2$ and note that $X_1 + X_2 + X_3 = 140$ have been found.

Example 2.8. This example illustrates Poissonization—the fact that some combinatorial probability problems become much easier when the number of objects is not fixed but has a Poisson distribution. Suppose that a Poisson number of Duke students with mean 2263 will show up to watch the next women's basketball game. What is the probability that for all of the 365 days there is at least one person in the crowd who has that birthday. (Pretend February 29th does not exist.)

By thinning if we let N_j be the number of people who have birthdays on the jth day of the year then the N_j are independent Poisson mean $2263/365 = 6.2$. The probability that all of $N_j > 0$ is

$$(1 - e^{-6.2})^{365} = 0.4764$$

The thinning results generalize easily to the nonhomogeneous case:

Theorem 2.12. *Suppose that in a Poisson process with rate λ, we keep a point that lands at s with probability $p(s)$. Then the result is a nonhomogeneous Poisson process with rate $\lambda p(s)$.*

For an application of this consider

Example 2.9 (M/G/∞ Queue). As one walks around the Duke campus it seems that every student is talking on their smartphone. The argument for arrivals at the ATM implies that the beginnings of calls follow a Poisson process. As for the calls themselves, while many people on the telephone show a lack of memory, there is no reason to suppose that the duration of a call has an exponential distribution, so we use a general distribution function G with $G(0) = 0$ and mean μ. Suppose that the system starts empty at time 0. The probability a call started at s has ended by time t is $G(t - s)$, so using Theorem 2.12 the number of calls still in progress at time t is Poisson with mean

$$\int_{s=0}^{t} \lambda(1 - G(t - s)) \, ds = \lambda \int_{r=0}^{t} (1 - G(r)) \, dr$$

Letting $t \to \infty$ we see that

Theorem 2.13. *In the long run the number of calls in the system will be Poisson with mean*

$$\lambda \int_{r=0}^{\infty} (1 - G(r)) \, dr = \lambda \mu$$

where in the second equality we have used (A.22). That is, the mean number in the system is the rate at which calls enter times their average duration. In the argument above we supposed that the system starts empty. Since the number of initial calls still in the system at time t decreases to 0 as $t \to \infty$, the limiting result is true for any initial number of calls X_0.

Example 2.10. Customers arrive at a sporting goods store at rate 10 per hour. 60 % of the customers are men and 40 % are women. Mean stay in the store for an amount of time that is exponential with mean 1/2 hour. Women for an amount of time that is uniformly distributed. What is the probability in equilibrium that there are four men and two women in the store?

By Poisson thinning the arrivals of men and women are independent Poisson process with rate 6 and 4. Since the mean time in the store is 1/2 for men and 1/4 for women, by Theorem 2.13 the number of men M and women W in equilibrium are independent Poissons with means 3 and 1. Thus

$$P(M = 4, W = 2) = e^{-3} \frac{3^4}{4!} \cdot e^{-1} \frac{1^2}{2!}$$

Example 2.11. People arrive at a puzzle exhibit according to a Poisson process with rate 2 per minute. The exhibit has enough copies of the puzzle so everyone at the exhibit can have one to play with. Suppose the puzzle takes an amount of time to solve that is uniform on $(0, 10)$ minutes. (a) What is the distribution of the number of people working on the puzzle in equilibrium? (b) What is the probability that there are three people at the exhibit working on puzzles, one that has been working more than four minutes, and two less than four minutes?

(a) The probability a customer who arrived x minutes ago is still working on the puzzle is $x/10$, so by Poisson thinning the number is Poisson with mean $2 \int_0^{10} x/10 \, dx = 10$. (b) The number that has been working more than four minutes is Poisson with mean $2 \int_0^6 x/10 \, dx = 36/10 = 3.6$, so the number less than four minutes is Poisson(6.4) and the answer is

$$e^{-3.6} \cdot 3.6 \cdot e^{-6.4} \cdot \frac{(6.4)^2}{2!}$$

2.4.2   Superposition

Taking one Poisson process and splitting it into two or more by using an i.i.d. sequence Y_i is called **thinning**. Going in the other direction and adding up a lot of independent processes is called **superposition**. Since a Poisson process can be split into independent Poisson processes, it should not be too surprising that when the independent Poisson processes are put together, the sum is Poisson with a rate equal to the sum of the rates.

Theorem 2.14. *Suppose $N_1(t), \ldots N_k(t)$ are independent Poisson processes with rates $\lambda_1, \ldots, \lambda_k$, then $N_1(t) + \cdots + N_k(t)$ is a Poisson process with rate $\lambda_1 + \cdots + \lambda_k$.*

Proof. It is clear that the sum has independent increments and $N_1(0) + N_2(0) = 0$. The fact that the increments have the right Poisson distribution follows from Theorem 2.4.   □

We will see in the next chapter that the ideas of compounding and thinning are very useful in computer simulations of continuous time Markov chains. For the moment we will illustrate their use in computing the outcome of races between Poisson processes.

Example 2.12 (A Poisson Race). Given a Poisson process of red arrivals with rate λ and an independent Poisson process of green arrivals with rate μ, what is the probability that we will get 6 red arrivals before a total of 4 green ones?

Solution. The first step is to note that the event in question is equivalent to having at least 6 red arrivals in the first 9. If this happens, then we have at most 3 green arrivals before the 6th red one. On the other hand, if there are 5 or fewer red arrivals in the first 9, then we have had at least 4 red arrivals and at most 5 green.

Viewing the red and green Poisson processes as being constructed by starting with one rate $\lambda + \mu$ Poisson process and flipping coins with probability $p = \lambda/(\lambda + \mu)$ to decide the color, we see that the probability of interest is

$$\sum_{k=6}^{9} \binom{9}{k} p^k (1-p)^{9-k}$$

If we suppose for simplicity that $\lambda = \mu$ so $p = 1/2$, this expression becomes

$$\frac{1}{512} \cdot \sum_{k=6}^{9} \binom{9}{k} = \frac{1 + 9 + (9 \cdot 8)/2 + (9 \cdot 8 \cdot 7)/3!}{512} = \frac{140}{512} = 0.273$$

2.4.3 Conditioning

Let $T_1, T_2, T_3, \ldots$ be the arrival times of a Poisson process with rate λ, let $U_1, U_2, \ldots U_n$ be independent and uniformly distributed on $[0, t]$, and let $V_1 < \ldots V_n$ be the U_i rearranged into increasing order . This section is devoted to the proof of the following remarkable fact.

Theorem 2.15. *If we condition on* $N(t) = n$, *then the vector* $(T_1, T_2, \ldots T_n)$ *has the same distribution as* $(V_1, V_2, \ldots V_n)$ *and hence the set of arrival times* $\{T_1, T_2, \ldots, T_n\}$ *has the same distribution as* $\{U_1, U_2, \ldots, U_n\}$.

Why is this true? We begin by finding the joint density function of (T_1, T_2, T_3) given that there were 3 arrivals before time t. The probability is 0 unless $0 < v_1 < v_2 < v_3 < t$. To compute the answer in this case, we note that $P(N(t) = 4) = e^{-\lambda t}(\lambda t)^3/3!$, and in order to have $T_1 = t_1, T_2 = t_2, T_3 = t_3, N(t) = 4$ we must have $\tau_1 = t_1, \tau_2 = t_2 - t_1, \tau_3 = t_3 - t_2$, and $\tau > t - t_3$, so the desired conditional distribution is

$$= \frac{\lambda e^{-\lambda t_1} \cdot \lambda e^{-\lambda(t_2 - t_1)} \cdot \lambda e^{-\lambda(t_3 - t_2)} \cdot e^{-\lambda(t - t_3)}}{e^{-\lambda t}(\lambda t)^3/3!}$$

$$= \frac{\lambda^3 e^{-\lambda t}}{e^{-\lambda t}(\lambda t)^3/3!} = \frac{3!}{t^3}$$

Note that the answer does not depend on the values of v_1, v_2, v_3 (as long as $0 < v_1 < v_2 < v_3 < t$), so the resulting conditional distribution is uniform over

$$\{(v_1, v_2, v_3) : 0 < v_1 < v_2 < v_3 < t\}$$

This set has volume $t^3/3!$ since $\{(v_1, v_2, v_3) : 0 < v_1, v_2, v_3 < t\}$ has volume t^3 and $v_1 < v_2 < v_3$ is one of 3! possible orderings.

Generalizing from the concrete example it is easy to see that the joint density function of $(T_1, T_2, \ldots T_n)$ given that there were n arrivals before time t is $n!/t^n$ for all times $0 < t_1 < \ldots < t_n < t$, which is the joint distribution of $(V_1, \ldots, V_n)$. The second fact follows easily from this, since there are $n!$ sets $\{T_1, T_2, \ldots T_n\}$ or $\{U_1, U_2, \ldots U_n\}$ for each ordered vector $(T_1, T_2, \ldots T_n)$ or $(V_1, V_2, \ldots, V_n)$. $\square$

Theorem 2.15 implies that if we condition on having n arrivals at time t, then the locations of the arrivals are the same as the location of n points thrown uniformly on $[0, t]$. From the last observation we immediately get

Theorem 2.16. *If* $s < t$ *and* $0 \le m \le n$, *then*

$$P(N(s) = m | N(t) = n) = \binom{n}{m} \left(\frac{s}{t}\right)^m \left(1 - \frac{s}{t}\right)^{n-m}$$

That is, the conditional distribution of $N(s)$ *given* $N(t) = n$ *is binomial*$(n, s/t)$.

Note that the answer does not depend on λ.

Proof. The number of arrivals by time s is the same as the number of $U_i < s$. The events $\{U_i < s\}$ these events are independent and have probability s/t, so the number of $U_i < s$ will be binomial$(n, s/t)$. □

One can also prove this directly.

Second Proof. By the definitions of conditional probability and of the Poisson process.

$$P(N(s) = m | N(t) = n) = \frac{P(N(s) = m)P(N(t) - N(s) = n - m)}{P(N(t) = n)}$$

$$= \frac{e^{-\lambda s}(\lambda s)^m/m! \cdot e^{-\lambda(t-s)}(\lambda(t-s))^{(n-m)}/(n-m)!}{e^{-\lambda t}(\lambda t)^n/n!}$$

Cancelling out the terms involving λ and rearranging the above

$$= \frac{n!}{m!(n-m)!} \left(\frac{s}{t}\right)^m \left(\frac{t-s}{t}\right)^{n-m}$$

which gives the desired result. □

Example 2.13. For a concrete example, suppose $N(3) = 4$.

$$P(N(1) = 1 | N(3) = 4) = 4 \cdot (1/3)^1 (2/3)^3 = 32/81$$

Similar calculations give the entire conditional distribution

k	0	1	2	3	4
$P(N(1) = k | N(3) = 4)$	16/81	32/81	24/81	8/81	1/81

Our final example provides a review of ideas from this section.

Example 2.14. Trucks and cars on highway US 421 are Poisson processes with rate 40 and 100 per hour, respectively. 1/8 of the trucks and 1/10 of the cars get off on exit 257 to go to the Bojangle's in Yadkinville. (a) Find the probability that exactly six trucks arrive at Bojangle's between noon and 1 p.m. (b) Given that there were six truck arrivals at Bojangle's between noon and 1 p.m., what is the probability that exactly two arrived between 12:20 and 12:40? (c) If we start watching at noon, what is the probability that four cars arrive before two trucks do. (d) Suppose that all trucks have 1 passenger while 30 % of the cars have 1 passenger, 50 % have 2, and 20 % have 4. Find the mean and standard deviation of the number of customers are that arrive at Bojangles' in one hour.

(a) By thinning trucks are Poisson with rate 5, so $e^{-5}5^6/6! = 0.1462$.
(b) By conditioning the probability is $C_{6,2}(1/3)^2(2/3)^4 = 0.3292$.

(c) For four cars to arrive before two trucks do, at least four of the first five arrivals
     must be cars. Trucks and cars are independent Poissons with rate 5 and 10 so
     the answer is

$$(2/3)^5 + 5(2/3)^4(1/3) = (7/3)(2/3)^4 = 0.4069$$

(d) The mean number of customers is

$$5 \cdot 1 + 10 \cdot [(0.3)1 + (0.5)2 + (0.2)4] = 26$$

The variance is

$$5 \cdot 1 + 10 \cdot [(0.3)1 + (0.5)4 + (0.2)16] = 55$$

so the standard deviation is $\sqrt{60} = 7.746$.

2.5   Chapter Summary

A random variable T is said to have **an exponential distribution with rate** λ, or
$T = \text{exponential}(\lambda)$, if $P(T \leq t) = 1 - e^{-\lambda t}$ for all $t \geq 0$. The mean is $1/\lambda$, variance
$1/\lambda^2$. The density function is $f_T(t) = \lambda e^{-\lambda t}$. The sum of n independent exponentials
has the gamma(n, λ) density

$$\lambda e^{-\lambda t} \frac{(\lambda t)^{n-1}}{(n-1)!}$$

Lack of Memory Property "if we've been waiting for t units of time then the
probability we must wait s more units of time is the same as if we haven't waited at
all."

$$P(T > t + s | T > t) = P(T > s)$$

Exponential Races Let $T_1, \ldots, T_n$ are independent, $T_i = \text{exponential}(\lambda_i)$, and $S = \min(T_1, \ldots, T_n)$. Then $S = \text{exponential}(\lambda_1 + \cdots + \lambda_n)$

$$P(T_i = \min(T_1, \ldots, T_n)) = \frac{\lambda_i}{\lambda_1 + \cdots + \lambda_n}$$

$\max\{S, T\} = S + T - \min\{S, T\}$ so taking expected value if $S = \text{exponential}(\mu)$
and $T = \text{exponential}(\lambda)$ then

$$E \max\{S, T\} = \frac{1}{\mu} + \frac{1}{\lambda} - \frac{1}{\mu + \lambda}$$

$$= \frac{1}{\mu + \lambda} + \frac{\lambda}{\lambda + \mu} \cdot \frac{1}{\mu} + \frac{\mu}{\lambda + \mu} \cdot \frac{1}{\lambda}$$

Poisson(μ) Distribution $P(X = n) = e^{-\mu}\mu^n/n!$. The mean and variance of X are μ.

Poisson Process Let $t_1, t_2, \ldots$ be independent exponential(λ) random variables. Let $T_n = t_1 + \ldots + t_n$ be the time of the nth arrival. Let $N(t) = \max\{n : T_n \le t\}$ be the number of arrivals by time t, which is Poisson(λt). $N(t)$ has **independent increments**: if $t_0 < t_1 < \ldots < t_n$, then $N(t_1) - N(t_0), N(t_2) - N(t_1), \ldots N(t_n) - N(t_{n-1})$ are independent.

Thinning Suppose we embellish our Poisson process by associating to each arrival an independent and identically distributed (i.i.d.) positive integer random variable Y_i. If we let $p_k = P(Y_i = k)$ and let $N_k(t)$ be the number of $i \le N(t)$ with $Y_i = k$, then $N_1(t), N_2(t), \ldots$ are independent Poisson processes and $N_k(t)$ has rate λp_k.

Random Sums Let $Y_1, Y_2, \ldots$ be i.i.d., let N be an independent nonnegative integer valued random variable, and let $S = Y_1 + \cdots + Y_N$ with $S = 0$ when $N = 0$.

 (i) If $E|Y_i|, EN < \infty$, then $ES = EN \cdot EY_i$.
 (ii) If $EY_i^2, EN^2 < \infty$, then $\text{var}(S) = EN \text{ var}(Y_i) + \text{var}(N)(EY_i)^2$.
 (iii) If N is Poisson(λ), $\text{var}(S) = \lambda E(Y_i^2)$

Superposition If $N_1(t)$ and $N_2(t)$ are independent Poison processes with rates λ_1 and λ_2, then $N_1(t) + N_2(t)$ is Poisson rate $\lambda_1 + \lambda_2$.

Conditioning Let $T_1, T_2, T_3, \ldots$ be the arrival times of a Poisson process with rate λ, and let $U_1, U_2, \ldots U_n$ be independent and uniformly distributed on $[0, t]$. If we condition on $N(t) = n$, then the set $\{T_1, T_2, \ldots T_n\}$ has the same distribution as $\{U_1, U_2, \ldots, U_n\}$.

2.6   Exercises

Exponential Distribution

2.1. Suppose that the time to repair a machine is exponentially distributed random variable with mean 2. (a) What is the probability the repair takes more than two hours. (b) What is the probability that the repair takes more than five hours given that it takes more than three hours.

2.2. The lifetime of a radio is exponentially distributed with mean 5 years. If Ted buys a 7-year-old radio, what is the probability it will be working 3 years later?

2.3. A doctor has appointments at 9 and 9:30. The amount of time each appointment lasts is exponential with mean 30. What is the expected amount of time after 9:30 until the second patient has completed his appointment?

2.4. Three people are fishing and each catches fish at rate 2 per hour. How long do we have to wait until everyone has caught at least one fish?

2.5. Ilan and Justin are competing in a math competition. They work independently and each has the same two problems to solve. The two problems take an exponentially distributed amount of time with mean 20 and 30 minutes respectively (or rates 3 and 2 if written in terms of hours). (a) What is the probability Ilan finishes both problems before Justin has completed the first one. (b) What is the expected time until both are done?

2.6. In a hardware store you must first go to server 1 to get your goods and then go to a server 2 to pay for them. Suppose that the times for the two activities are exponentially distributed with means six and three minutes. Compute the average amount of time it takes Bob to get his goods and pay if when he comes in there is one customer named Al with server 1 and no one at server 2.

2.7. Consider a bank with two tellers. Three people, Anne, Betty, and Carol enter the bank at almost the same time and in that order. Anne and Betty go directly into service while Carol waits for the first available teller. Suppose that the service times for two servers are exponentially distributed with mean three and six minutes (or they have rates of 20 and 10 per hour). (a) What is the expected total amount of time for Carol to complete her businesses? (b) What is the expected total time until the last of the three customers leaves? (c) What is the probability for Anne, Betty, and Carol to be the last one to leave?

2.8. A flashlight needs two batteries to be operational. You start with four batteries numbered 1–4. Whenever a battery fails it is replaced by the lowest-numbered working battery. Suppose that battery life is exponential with mean 100 hours. Let T be the time at which there is one working battery left and N be the number of the one battery that is still good. (a) Find ET. (b) Find the distribution of N.

2.9. Excited by the recent warm weather Jill and Kelly are doing spring cleaning at their apartment. Jill takes an exponentially distributed amount of time with mean 30 minutes to clean the kitchen. Kelly takes an exponentially distributed amount of time with mean 40 minutes to clean the bathroom. The first one to complete their task will go outside and start raking leaves, a task that takes an exponentially distributed amount of time with a mean of one hour. When the second person is done inside, they will help the other and raking will be done at rate 2. (Of course the other person may already be done raking in which case the chores are done.) What is the expected time until the chores are all done?

2.10. Ron, Sue, and Ted arrive at the beginning of a professor's office hours. The amount of time they will stay is exponentially distributed with means of 1, 1/2, and 1/3 hour. (a) What is the expected time until only one student remains? (b) For each student find the probability they are the last student left. (c) What is the expected time until all three students are gone?

Poisson Approximation to Binomial

2.11. Compare the Poisson approximation with the exact binomial probabilities of 1 success when $n = 20, p = 0.1$.

2.12. Compare the Poisson approximation with the exact binomial probabilities of no success when (a) $n = 10, p = 0.1$, (b) $n = 50, p = 0.02$.

2.13. The probability of a three of a kind in poker is approximately 1/50. Use the Poisson approximation to estimate the probability you will get at least one three of a kind if you play 20 hands of poker.

2.14. Suppose 1 % of a certain brand of Christmas lights is defective. Use the Poisson approximation to compute the probability that in a box of 25 there will be at most one defective bulb.

Poisson Processes: Basic Properties

2.15. Suppose $N(t)$ is a Poisson process with rate 3. Let T_n denote the time of the nth arrival. Find (a) $E(T_{12})$, (b) $E(T_{12}|N(2) = 5)$, (c) $E(N(5)|N(2) = 5)$.

2.16. Customers arrive at a shipping office at times of a Poisson process with rate 3 per hour. (a) The office was supposed to open at 8 a.m. but the clerk Oscar overslept and came in at 10 a.m. What is the probability that no customers came in the two-hour period? (b) What is the distribution of the amount of time Oscar has to wait until his first customer arrives?

2.17. Suppose that the number of calls per hour to an answering service follows a Poisson process with rate 4. (a) What is the probability that fewer (i.e., <) than 2 calls came in the first hour? (b) Suppose that 6 calls arrive in the first hour, what is the probability there will be < 2 in the second hour. (c) Suppose that the operator gets to take a break after she has answered 10 calls. How long are her average work periods?

2.18. Traffic on Rosedale Road in Princeton, NJ, follows a Poisson process with rate 6 cars per minute. A deer runs out of the woods and tries to cross the road. If there is a car passing in the next five seconds, then there will be a collision. (a) Find the probability of a collision. (b) What is the chance of a collision if the deer only needs two seconds to cross the road.

2.19. Calls to the Dryden fire department arrive according to a Poisson process with rate 0.5 per hour. Suppose that the time required to respond to a call, return to the station, and get ready to respond to the next call is uniformly distributed between 1/2 and 1 hour. If a new call comes before the Dryden fire department is ready to respond, the Ithaca fire department is asked to respond. Suppose that the Dryden fire department is ready to respond now. Find the probability distribution for the number of calls they will handle before they have to ask for help from the Ithaca fire department.

2.20. A math professor waits at the bus stop at the Mittag-Leffler Institute in the suburbs of Stockholm, Sweden. Since he has forgotten to find out about the bus schedule, his waiting time until the next bus is uniform on $(0,1)$. Cars drive by the bus stop at rate 6 per hour. Each will take him into town with probability $1/3$. What is the probability he will end up riding the bus?

2.21. The number of hours between successive trains is T which is uniformly distributed between 1 and 2. Passengers arrive at the station according to a Poisson process with rate 24 per hour. Let X denote the number of people who get on a train. Find (a) EX, (b) var (X).

2.22. Let T be exponentially distributed with rate λ. (a) Use the definition of conditional expectation to compute $E(T|T < c)$. (b) Determine $E(T|T < c)$ from the identity

$$ET = P(T < c)E(T|T < c) + P(T > c)E(T|T > c)$$

2.23. *When did the chicken cross the road?* Suppose that traffic on a road follows a Poisson process with rate λ cars per minute. A chicken needs a gap of length at least c minutes in the traffic to cross the road. To compute the time the chicken will have to wait to cross the road, let $t_1, t_2, t_3, \ldots$ be the interarrival times for the cars and let $J = \min\{j : t_j > c\}$. If $T_n = t_1 + \cdots + t_n$, then the chicken will start to cross the road at time T_{J-1} and complete his journey at time $T_{J-1} + c$. Use the previous exercise to show $E(T_{J-1} + c) = (e^{\lambda c} - 1)/\lambda$.

Random Sums

2.24. Edwin catches trout at times of a Poisson process with rate 3 per hour. Suppose that the trout weigh an average of four pounds with a standard deviation of two pounds. Find the mean and standard deviation of the total weight of fish he catches in two hours.

2.25. An insurance company pays out claims at times of a Poisson process with rate 4 per week. Writing K as shorthand for "thousands of dollars," suppose that the mean payment is 10K and the standard deviation is 6K. Find the mean and standard deviation of the total payments for 4 weeks.

2.26. Customers arrive at an automated teller machine at the times of a Poisson process with rate of 10 per hour. Suppose that the amount of money withdrawn on each transaction has a mean of $30 and a standard deviation of $20. Find the mean and standard deviation of the total withdrawals in eight hours.

Thinning and Conditioning

2.27. Rock concert tickets are sold at a ticket counter. Females and males arrive at times of independent Poisson processes with rates 30 and 20 customers per hour. (a) What is the probability the first three customers are female? (b) If exactly two customers arrived in the first five minutes, what is the probability both arrived in the first three minutes. (c) Suppose that customers regardless of sex buy one ticket with probability 1/2, two tickets with probability 2/5, and three tickets with probability 1/10. Let N_i be the number of customers that buy i tickets in the first hour. Find the joint distribution of (N_1, N_2, N_3).

2.28. A light bulb has a lifetime that is exponential with a mean of 200 days. When it burns out a janitor replaces it immediately. In addition there is a handyman who comes at times of a Poisson process at rate .01 and replaces the bulb as "preventive maintenance." (a) How often is the bulb replaced? (b) In the long run what fraction of the replacements are due to failure?

2.29. Calls originate from Dryden according to a rate 12 Poisson process. 3/4 are local and 1/4 are long distance. Local calls last an average of ten minutes, while long distance calls last an average of five minutes. Let M be the number of local calls and N the number of long distance calls in equilibrium. Find the distribution of (M, N). what is the number of people on the line.

2.30. Suppose that the number of calls per hour to an answering service follows a Poisson process with rate 4. Suppose that 3/4's of the calls are made by men, 1/4 by women, and the sex of the caller is independent of the time of the call. (a) What is the probability that in one hour exactly two men and three women will call the answering service? (b) What is the probability 3 men will make phone calls before three women do?

2.31. Suppose $N(t)$ is a Poisson process with rate 2. Compute the conditional probabilities (a) $P(N(3) = 4|N(1) = 1)$, (b) $P(N(1) = 1|N(3) = 4)$.

2.32. For a Poisson process $N(t)$ with arrival rate 2 compute: (a) $P(N(2) = 5)$, (b) $P(N(5) = 8|N(2) = 3)$, (c) $P(N(2) = 3|N(5) = 8)$.

2.33. Customers arrive at a bank according to a Poisson process with rate 10 per hour. Given that two customers arrived in the first five minutes, what is the probability that (a) both arrived in the first two minutes. (b) at least one arrived in the first two minutes.

2.34. Wayne Gretsky scored a Poisson mean 6 number of points per game. 60 % of these were goals and 40 % were assists (each is worth one point). Suppose he is paid a bonus of 3K for a goal and 1K for an assist. (a) Find the mean and standard deviation for the total revenue he earns per game. (b) What is the probability that he has four goals and two assists in one game? (c) Conditional on the fact that he had six points in a game, what is the probability he had 4 in the first half?

2.35. Hockey teams 1 and 2 score goals at times of Poisson processes with rates 1 and 2. Suppose that $N_1(0) = 3$ and $N_2(0) = 1$. (a) What is the probability that $N_1(t)$ will reach 5 before $N_2(t)$ does? (b) Answer part (a) for Poisson processes with rates λ_1 and λ_2.

2.36. Traffic on Snyder Hill Road in Ithaca, NY, follows a Poisson process with rate 2/3's of a vehicle per minute. 10 % of the vehicles are trucks, the other 90 % are cars. (a) What is the probability at least one truck passes in a hour? (b) Given that ten trucks have passed by in an hour, what is the expected number of vehicles that have passed by. (c) Given that 50 vehicles have passed by in a hour, what is the probability there were exactly 5 trucks and 45 cars.

2.37. As a community service members of the Mu Alpha Theta fraternity are going to pick up cans from along a roadway. A Poisson mean 60 members show up for work. 2/3 of the workers are enthusiastic and will pick up a mean of ten cans with a standard deviation of 5. 1/3 of the workers are lazy and will only pick an average of three cans with a standard deviation of 2. Find the mean and standard deviation of the number of cans collected.

2.38. Suppose that Virginia scores touchdowns at rate λ_7 and field goals at rate λ_3 while Duke scores touchdowns at rate μ_7 and field goals at rate μ_3. The subscripts indicate the number of points the team receives for each type of event. The final score in the Virginia-Duke game in 2015 was 42–34, i.e., Virginia scored six touchdowns while Duke scored four touchdowns and two field goals. Given this outcome what is the probability that the score at halftime was 28–13, i.e., four touchdowns for Virginia, versus one touchdown and two field goals for Duke. Write your answer as a decimal, e.g., 0.12345.

2.39. A Philadelphia taxi driver gets new customers at rate 1/5 per minute. With probability 1/3 the person wants to go to the airport, a 20-minute trip. After waiting in a line of cabs at the airport for an average of 35 minutes, he gets another fare and spends 20 minutes driving back to drop that person off. Each trip to or from the airport costs the customer a standard $28 charge (ignore tipping). While in the city fares with probability 2/3 want a short trip that lasts an amount of time uniformly distributed on [2, 10] minutes and the cab driver earns an average of $1.33 a minute (i.e., 4/3 of a dollar). (a) In the long run how much money does he make per hour? (b) What fraction of time does he spend going to and from the airport (inducing the time spent in line there)?

2.40. People arrive at the Durham Farmer's market at rate 15 per hour. 4/5's are vegetarians, and 1/5 are meat eaters. Vegetarians spend an average of $7 with a standard deviation of 3. Meat eaters spend an average of $15 with a standard deviation of 8. (a) Compute the probability that in the first 20 minutes exactly three vegetarians and two meat eaters arrive. You do not have to simplify your answer. (b) Find the mean and standard deviation of the amount of money spent during the four hours the market is open.

2.41. Vehicles carrying **Occupy Durham** protesters arrive at rate 30 per hour.

- 50 % are bicycles carrying one person
- 30 % are BMW's carrying two people
- 20 % are Prius's carrying four happy carpoolers

(a) What is the probability exactly 4 Prius's arrive between 12 and 12:30? (b) Find the mean and standard deviation of the number of people which arrive in that half hour.

2.42. People arrive at the Southpoint Mall at times of a Poisson process with rate 96 per hour. 1/3 of the shoppers are men and 2/3 are women. (a) Women shop for an amount of time that is exponentially distributed with mean three hours. Men shop for a time that is uniformly distributed on $[0, 1]$ hour. Let (M_t, W_t) be the number of men and women at time t. What is the equilibrium distribution for (M_t, W_t).

(b) Women spend an average of $150 with a standard deviation of $80. Find the mean and variance of the amount of money spent by women who arrived between 9 a.m. and 10 a.m.

2.43. A policewoman on the evening shift writes a Poisson mean 6 number of tickets per hour. 2/3's of these are for speeding and cost $100. 1/3's of these are for DWI and cost $400. (a) Find the mean and standard deviation for the total revenue from the tickets she writes in an hour. (b) What is the probability that between 2 a.m. and 3 a.m. she writes five tickets for speeding and one for DWI. (c) Let A be the event that she writes no tickets between 1 a.m. and 1:30, and N be the number of tickets she writes between 1 a.m. and 2 a.m. Which is larger $P(A)$ or $P(A|N = 5)$? Don't just answer yes or no, compute both probabilities.

2.44. Ignoring the fact that the bar exam is only given twice a year, let us suppose that new lawyers arrive in Los Angeles according to a Poisson process with mean 300 per year. Suppose that each lawyer independently practices for an amount of time T with a distribution function $F(t) = P(T \le t)$ that has $F(0) = 0$ and mean 25 years. Show that in the long run the number of lawyers in Los Angeles is Poisson with mean 7500.

More Theoretical Exercises

2.45. Copy machine 1 is in use now. Machine 2 will be turned on at time t. Suppose that the machines fail at rate λ_i. What is the probability that machine 2 is the first to fail?

2.46. Customers arrive according to a Poisson process of rate λ per hour. Joe does not want to stay until the store closes at $T = 10$ p.m., so he decides to close up when the first customer after time $T - s$ arrives. He wants to leave early but he does not want to lose any business so he is happy if he leaves before T and no one arrives after. (a) What is the probability he achieves his goal? (b) What is the optimal value of s and the corresponding success probability?

2.47. Let S and T be exponentially distributed with rates λ and μ. Let $U = \min\{S, T\}$ and $V = \max\{S, T\}$. Find (a) EU. (b) $E(V - U)$, (c) EV. (d) Use the identity $V = S + T - U$ to get a different looking formula for EV and verify the two are equal.

2.48. Let S and T be exponentially distributed with rates λ and μ. Let $U = \min\{S, T\}$, $V = \max\{S, T\}$, and $W = V - U$. Find the variances of U, V, and W.

2.49. Consider a bank with two tellers. Three people, Alice, Betty, and Carol enter the bank at almost the same time and in that order. Alice and Betty go directly into service while Carol waits for the first available teller. Suppose that the service times for each teller are exponentially distributed with rates $\lambda \leq \mu$. (a) What is the expected total amount of time for Carol to complete her businesses? (b) What is the expected total time until the last of the three customers leaves? (c) What is the probability Carol is the last one to leave?

2.50. A flashlight needs two batteries to be operational. You start with n batteries numbered 1 to n. Whenever a battery fails it is replaced by the lowest-numbered working battery. Suppose that battery life is exponential with mean 100 hours. Let T be the time at which there is one working battery left and N be the number of the one battery that is still good. (a) Find ET. (b) Find the distribution of N.

2.51. Let T_i, $i = 1, 2, 3$ be independent exponentials with rate λ_i. (a) Show that for any numbers t_1, t_2, t_3

$$\max\{t_1, t_2, t_3\} = t_1 + t_2 + t_3 - \min\{t_1, t_2\} - \min\{t_1, t_3\}$$
$$- \min\{t_2, t_3\} + \min\{t_1, t_2, t_3\}$$

(b) Use (a) to find $E \max\{T_1, T_2, T_3\}$. (c) Use the formula to give a simple solution of part (c) of Exercise 2.10.

2.52. Consider a Poisson process with rate λ and let L be the time of the last arrival in the interval $[0, t]$, with $L = 0$ if there was no arrival. (a) Compute $E(t - L)$ (b) What happens when we let $t \to \infty$ in the answer to (a)?

2.53. Policy holders of an insurance company have accidents at times of a Poisson process with rate λ. The distribution of the time R until a claim is reported is random with $P(R \leq r) = G(r)$ and $ER = \nu$. (a) Find the distribution of the number of unreported claims. (b) Suppose each claim has mean μ and variance σ^2. Find the mean and variance of S the total size of the unreported claims.

2.54. Let S_t be the price of stock at time t and suppose that at times of a Poisson process with rate λ the price is multiplied by a random variable $X_i > 0$ with mean μ and variance σ^2. That is,

$$S_t = S_0 \prod_{i=1}^{N(t)} X_i$$

where the product is 1 if $N(t) = 0$. Find $ES(t)$ and $\text{var}\, S(t)$.

2.55. Let $\{N(t), t \geq 0\}$ be a Poisson process with rate λ. Let $T \geq 0$ be an independent with mean μ and variance σ^2. Find cov (T, N_T).

2.56. Messages arrive to be transmitted across the internet at times of a Poisson process with rate λ. Let Y_i be the size of the ith message, measured in bytes, and let $g(z) = Ez^{Y_i}$ be the generating function of Y_i. Let $N(t)$ be the number of arrivals at time t and $S = Y_1 + \cdots + Y_{N(t)}$ be the total size of the messages up to time t. (a) Find the generating function $f(z) = E(z^S)$. (b) Differentiate and set $z = 1$ to find ES. (c) Differentiate again and set $z = 1$ to find $E\{S(S-1)\}$. (d) Compute var (S).

2.57. Consider a Poisson process with rate λ and let L be the time of the last arrival in the interval $[0, t]$, with $L = 0$ if there was no arrival. (a) Compute $E(t - L)$ (b) What happens when we let $t \to \infty$ in the answer to (a)?

2.58. Let $t_1, t_2, \ldots$ be independent exponential(λ) random variables and let N be an independent random variable with $P(N = n) = (1 - p)^{n-1}$. What is the distribution of the random sum $T = t_1 + \cdots + t_N$?

2.59. Signals are transmitted according to a Poisson process with rate λ. Each signal is successfully transmitted with probability p and lost with probability $1 - p$. The fates of different signals are independent. For $t \geq 0$ let $N_1(t)$ be the number of signals successfully transmitted and let $N_2(t)$ be the number that are lost up to time t. (a) Find the distribution of $(N_1(t), N_2(t))$. (b) What is the distribution of $L =$ the number of signals lost before the first one is successfully transmitted?

2.60. Starting at some fixed time, which we will call 0 for convenience, satellites are launched at times of a Poisson process with rate λ. After an independent amount of time having distribution function F and mean μ, the satellite stops working. Let $X(t)$ be the number of working satellites at time t. (a) Find the distribution of $X(t)$. (b) Let $t \to \infty$ in (a) to show that the limiting distribution is Poisson$(\lambda\mu)$.

2.61. Consider two independent Poisson processes $N_1(t)$ and $N_2(t)$ with rates λ_1 and λ_2. What is the probability that the two-dimensional process $(N_1(t), N_2(t))$ ever visits the point (i, j)?

Chapter 3
Renewal Processes

3.1 Laws of Large Numbers

In the Poisson process the times between successive arrivals are independent and exponentially distributed. The lack of memory property of the exponential distribution is crucial for many of the special properties of the Poisson process derived in this chapter. However, in many situations the assumption of exponential interarrival times is not justified. In this section we will consider a generalization of Poisson processes called **renewal processes** in which the times $t_1, t_2, \ldots$ between events are independent and have distribution F.

In order to have a simple metaphor with which to discuss renewal processes, we will think of a single light bulb maintained by a very diligent janitor, who replaces the light bulb immediately after it burns out. Let t_i be the lifetime of the ith light bulb. We assume that the light bulbs are bought from one manufacturer, so we suppose

$$P(t_i \le t) = F(t)$$

where F is a distribution function with $F(0) = P(t_i \le 0) = 0$.

If we start with a new bulb (numbered 1) at time 0 and each light bulb is replaced when it burns out, then $T_n = t_1 + \cdots + t_n$ gives the time that the nth bulb burns out, and

$$N(t) = \max\{n : T_n \le t\}$$

is the number of light bulbs that have been replaced by time t. The picture is the same as the one for the Poisson process:

© Springer International Publishing Switzerland 2016

R. Durrett, *Essentials of Stochastic Processes*, Springer Texts in Statistics,

DOI 10.1007/978-3-319-45614-0_3

If renewal theory were only about changing light bulbs, it would not be a very useful subject. The reason for our interest in this system is that it captures the essence of a number of different situations. On example that we have already seen is

Example 3.1 (Markov Chains). Let X_n be a Markov chain and suppose that $X_0 = x$. Let T_n be the nth time that the process returns to x. The strong Markov property implies that $t_n = T_n - T_{n-1}$ are independent, so T_n is a renewal process.

Example 3.2 (Machine Repair). Instead of a light bulb, think of a machine that works for an amount of time s_i before it fails, requiring an amount of time u_i to be repaired. Let $t_i = s_i + u_i$ be the length of the ith cycle of breakdown and repair. If we assume that the repair leaves the machine in a "like new" condition, then the t_i are independent and identically distributed (i.i.d.) and a renewal process results.

The first important result about renewal processes is the following law of large numbers:

Theorem 3.1. *Let* $\mu = Et_i$ *be mean interarrival time. If* $P(t_i > 0) > 0$, *then with probability one,*

$$N(t)/t \to 1/\mu \quad as\ t \to \infty$$

In words, this says that if our light bulb lasts μ years on the average then in t years we will use up about t/μ light bulbs. Since the interarrival times in a Poisson process are exponential with mean $1/\lambda$, Theorem 3.1 implies that if $N(t)$ is the number of arrivals up to time t in a Poisson process, then

$$N(t)/t \to \lambda \quad as\ t \to \infty \tag{3.1}$$

Proof of Theorem 3.1. We use the   □

Theorem 3.2 (Strong Law of Large Numbers). *Let* $x_1, x_2, x_3, \ldots$ *be i.i.d. with* $Ex_i = \mu$, *and let* $S_n = x_1 + \cdots + x_n$. *Then with probability one,*

$$S_n/n \to \mu \quad as\ n \to \infty$$

Taking $x_i = t_i$, we have $S_n = T_n$, so Theorem 3.2 implies that with probability one, $T_n/n \to \mu$ as $n \to \infty$. Now by definition,

$$T_{N(t)} \le t < T_{N(t)+1}$$

Dividing by $N(t)$, we have

$$\frac{T_{N(t)}}{N(t)} \le \frac{t}{N(t)} \le \frac{T_{N(t)+1}}{N(t)+1} \cdot \frac{N(t)+1}{N(t)}$$

By the strong law of large numbers, the left- and right-hand sides converge to μ. From this it follows that $t/N(t) \to \mu$ and hence $N(t)/t \to 1/\mu$.   □

Our next topic is a simple extension of the notion of a renewal process that greatly extends the class of possible applications. We suppose that at the time of the ith renewal we earn a reward r_i. The reward r_i may depend on the ith interarrival time t_i, but we will assume that the pairs (r_i, t_i), $i = 1, 2, \ldots$ are independent and have the same distribution. Let

$$R(t) = \sum_{i=1}^{N(t)} r_i$$

be the total amount of rewards earned by time t. The main result about renewal reward processes is the following strong law of large numbers.

Theorem 3.3. *With probability one,*

$$\frac{R(t)}{t} \to \frac{Er_i}{Et_i} \tag{3.2}$$

Proof. Multiplying and dividing by $N(t)$, we have

$$\frac{R(t)}{t} = \left(\frac{1}{N(t)} \sum_{i=1}^{N(t)} r_i \right) \frac{N(t)}{t} \to Er_i \cdot \frac{1}{Et_i}$$

where in the last step we have used Theorem 3.1 and applied the strong law of large numbers to the sequence r_i. Here and in what follows we are ignoring rewards earned in the interval $[T_{N(t)}, t]$. These do not effect the limit but proving this is not trivial.   □

Intuitively, (3.2) can be written as

$$\text{limiting reward/time} = \frac{\text{expected reward/cycle}}{\text{expected time/cycle}}$$

To illustrate the use of Theorem 3.3 we consider

Example 3.3 (Long Run Car Costs). Suppose that the lifetime of a car is a random variable with density function h. Our methodical Mr. Brown buys a new car as soon as the old one breaks down or reaches T years. Suppose that a new car costs A dollars and that an additional cost of B dollars to repair the vehicle is incurred if it breaks down before time T. What is the long-run cost per unit time of Mr. Brown's policy?

Solution. The duration of the ith cycle, t_i, has

$$Et_i = \int_0^T th(t)\, dt + T \int_T^\infty h(t)\, dt$$

since the length of the cycle will be t_i if the car's life is $t_i < T$, but T if the car's life $t_i \geq T$. The reward (or cost) of the ith cycle has

$$Er_i = A + B \int_0^T h(t)\, dt$$

since Mr. Brown always has to pay A dollars for a new car but only owes the additional B dollars if the car breaks down before time T. Using Theorem 3.3 we see that the long run cost per unit time is

$$\frac{Er_i}{Et_i} = \frac{A + B \int_0^T h(t)\, dt}{\int_0^T th(t)\, dt + \int_T^\infty Th(t)\, dt}$$

Concrete Example Suppose that the lifetime of Mr. Brown's car is uniformly distributed on $[0, 10]$. This is probably not a reasonable assumption, since when cars get older they have a greater tendency to break. However, it gives us a relatively simple concrete example. Suppose that the cost of a new car is $A = 10$ (thousand dollars), while the breakdown cost is $B = 3$ (thousand dollars). If Mr. Brown replaces his car after T years, then the expected values of interest are

$$Er_i = 10 + 3\frac{T}{10} = 10 + 0.3T$$

$$Et_i = \int_0^T \frac{t}{10}\, dt + T\left(1 - \frac{T}{10}\right) = \frac{T^2}{20} + T - \frac{T^2}{10} = T - 0.05T^2$$

Combining the expressions for the Er_i and Et_i we see that the long-run cost per unit time is

$$\frac{Er_i}{Et_i} = \frac{10 + 0.3T}{T - 0.05T^2}$$

To maximize we take the derivative

$$\frac{d}{dT}\frac{Er_i}{Et_i} = \frac{0.3(T - 0.05T^2) - (10 + 0.3T)(1 - 0.1T)}{(T - 0.1T^2)^2}$$

$$= \frac{0.3T - 0.015T^2 - 10 - 0.3T + T + 0.03T^2}{(T - 0.1T^2)^2}$$

The numerator is $0.015T^2 + T - 10$ which is 0 when

$$T = \frac{-1 \pm \sqrt{1 + 4(0.015)(10)}}{2(0.015)} = \frac{-1 \pm \sqrt{1.6}}{0.03}$$

We want the $+$ root which is $T = 8.83$ years.

Using the idea of renewal reward processes, we can easily treat the following extension of renewal processes.

Example 3.4 (Alternating Renewal Processes). Let $s_1, s_2, \ldots$ be independent with a distribution F that has mean μ_F, and let $u_1, u_2, \ldots$ be independent with distribution G that has mean μ_G. For a concrete example consider the machine in Example 3.2 that works for an amount of time s_i before needing a repair that takes u_i units of time. However, to talk about things in general we will say that the alternating renewal process spends an amount of time s_i in state 1, an amount of time u_i in state 2, and then repeats the cycle again.

Theorem 3.4. *In an alternating renewal process, the limiting fraction of time in state 1 is*

$$\frac{\mu_F}{\mu_F + \mu_G}$$

To see that this is reasonable and to help remember the formula, consider the nonrandom case. If the machine always works for exactly μ_F days and then needs repair for exactly μ_G days, then the limiting fraction of time spent working is $\mu_F/(\mu_F + \mu_G)$.

Proof. In order to compute the limiting fraction of time the machine is working we let $t_i = s_i + u_i$ be the duration of the ith cycle, and let the reward $r_i = s_i$, the amount of time the machine was working during the ith cycle. In this case, Theorem 3.3 implies that

$$\frac{R(t)}{t} \to \frac{Er_i}{Et_i} = \frac{\mu_F}{\mu_F + \mu_G}$$

which gives the desired result. □

For a concrete example of alternating renewal processes, consider

Example 3.5 (Poisson Janitor). A light bulb burns for an amount of time having distribution F with mean μ_F then burns out. A janitor comes at times of a rate λ Poisson process to check the bulb and will replace the bulb if it is burnt out. (a) At what rate are bulbs replaced? (b) What is the limiting fraction of time that the light bulb works? (c) What is the limiting fraction of visits on which the bulb is working?

Solution. Suppose that a new bulb is put in at time 0. It will last for an amount of time s_1. Using the lack of memory property of the exponential distribution, it follows that the amount of time until the next inspection, u_1, will have an exponential distribution with rate λ. The bulb is then replaced and the cycle starts again, so we have an alternating renewal process.

To answer (a), we note that the expected length of a cycle $Et_i = \mu_F + 1/\lambda$, so if $N(t)$ is the number of bulbs replaced by time t, then it follows from Theorem 3.1 that

$$\frac{N(t)}{t} \to \frac{1}{\mu_F + 1/\lambda}$$

In words, bulbs are replaced on the average every $\mu_F + 1/\lambda$ units of time.

To answer (b), we let $r_i = s_i$, so Theorem 3.4 implies that in the long run, the fraction of time the bulb has been working up to time t is

$$\frac{Er_i}{Et_i} = \frac{\mu_F}{\mu_F + 1/\lambda}$$

To answer (c), we note that if $V(t)$ is the number of visits the janitor has made by time t, then by the law of large numbers for the Poisson process we have

$$\frac{V(t)}{t} \to \lambda$$

Combining this with the result of (a), we see that the fraction of visits on which bulbs are replaced

$$\frac{N(t)}{V(t)} \to \frac{1/(\mu_F + 1/\lambda)}{\lambda} = \frac{1/\lambda}{\mu_F + 1/\lambda}$$

This answer is reasonable since it is also the limiting fraction of time the bulb is off.

3.2 Applications to Queueing Theory

In this section we will use the ideas of renewal theory to prove results for queueing systems with general service times. In the first part of this section we will consider general arrival times. In the second we will specialize to Poisson arrivals.

3.2.1 GI/G/1 Queue

Here the *GI* stands for general input. That is, we suppose that the times t_i between successive arrivals are independent and have a distribution F with mean $1/\lambda$. We make this somewhat unusual choice of notation for mean so that if $N(t)$ is the number of arrivals by time t, then Theorem 3.1 implies that the long-run arrival rate is

$$\lim_{t \to \infty} \frac{N(t)}{t} = \frac{1}{Et_i} = \lambda$$

The second G stands for general service times. That is, we assume that the ith customer requires an amount of service s_i, where the s_i are independent and have a distribution G with mean $1/\mu$. Again, the notation for the mean is chosen so that the service rate is μ. The final 1 indicates there is one server. Our first result states that the queue is stable if the arrival rate is smaller than the long-run service rate.

Theorem 3.5. *Suppose* $\lambda < \mu$. *If the queue starts with some finite number* $k \geq 1$ *customers who need service, then it will empty out with probability one. Furthermore, the limiting fraction of time the server is busy is* $\leq \lambda/\mu$.

Proof. Let $T_n = t_1 + \cdots + t_n$ be the time of the nth arrival. The strong law of large numbers, Theorem 3.2 implies that

$$\frac{T_n}{n} \to \frac{1}{\lambda}$$

Let Z_0 be the sum of the service times of the customers in the system at time 0 and let s_i be the service time of the ith customer to arrive after time 0, and let $S_n = s_1 + \cdots + s_n$. The strong law of large numbers implies

$$\frac{Z_0 + S_n}{n} \to \frac{1}{\mu}$$

The amount of time the server has been busy up to time T_n is $\leq Z_0 + S_n$. Using the two results $(Z_0 + S_n)/T_n \to \lambda/\mu$, which proves the desired result. □

Remark. The actual time spent working in $[0, T_n]$ is $Z_0 + S_n - Z_n$ where Z_n is the amount of work in the system at time T_n, i.e., the amount of time needed to empty the system if there were no more arrivals. To argue that equality holds we need to show that $Z_n/n \to 0$. Intuitively, the condition $\lambda < \mu$ implies that the chain is positive recurrent, so EZ_n stays bounded, and hence $Z_n/n \to 0$. Theorem 5.14 will show this. However, in Example 3.6 we will give a very simple proof that the limiting fraction of time the server is busy is λ/μ.

3.2.2 Cost Equations

In this subsection we will prove some general results about the GI/G/1 queue that come from very simple arguments. Let X_s be the number of customers in the system at time s. Let L be the long-run average number of customers in the system:

$$L = \lim_{t \to \infty} \frac{1}{t} \int_0^t X_s \, ds$$

Let W be the long-run average amount of time a customer spends in the system:

$$W = \lim_{n \to \infty} \frac{1}{n} \sum_{m=1}^n W_m$$

where W_m is the amount of time the mth arriving customer spends in the system. Finally, let λ_a be the long-run average rate at which arriving customers join the system, that is,

$$\lambda_a = \lim_{t\to\infty} N_a(t)/t$$

where $N_a(t)$ is the number of customers who arrive before time t and enter the system. Ignoring the problem of proving the existence of these limits, we can assert that these quantities are related by

Theorem 3.6 (Little's Formula). $L = \lambda_a W$.

Why is this true? Suppose each customer pays \$1 for each minute of time she is in the system. When ℓ customers are in the system, we are earning \$$\ell$ per minute, so in the long run we earn an average of \$$L$ per minute. On the other hand, if we imagine that customers pay for their entire waiting time when they arrive then we earn at rate $\lambda_a W$ per minute, i.e., the rate at which customers enter the system multiplied by the average amount they pay.   □

Example 3.6 (Waiting Time in the Queue). Consider the $GI/G/1$ queue and suppose that we are only interested in the customer's average waiting time in the queue, W_Q. If we know the average waiting time W in the system, this can be computed by simply subtracting out the amount of time the customer spends in service

$$W_Q = W - Es_i \tag{3.3}$$

Let L_Q be the average queue length in equilibrium; i.e., we do not count the customer in service if there is one. If suppose that customers pay \$1 per minute in the queue and repeat the derivation of Little's formula, then

$$L_Q = \lambda_a W_Q \tag{3.4}$$

The length of the queue is 1 less than the number in the system, except when the number in the system is 0, so if $\pi(0)$ is the probability of no customers, then

$$L_Q = L - 1 + \pi(0)$$

Combining the last three equations with our first cost equation:

$$\pi(0) = L_Q - (L - 1) = 1 + \lambda_a(W_Q - W) = 1 - \lambda_a Es_i$$

Recalling that $Es_i \overset{!}{=} 1/\mu$, we have

$$\pi(0) = 1 - \frac{\lambda_a}{\mu} \tag{3.5}$$

In the $GI/G/1$ queue $\lambda_a = \lambda$ so the inequality in Theorem 3.5 is sharp.

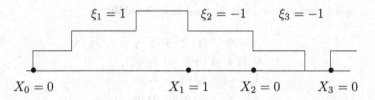

Fig. 3.1 Realization of the M/G/1 queue. *Black dots* indicate the times at which the customers enter service

3.2.3 M/G/1 Queue

Here the M stands for Markovian input and indicates we are considering the special case of the $GI/G/1$ queue in which the inputs are a rate λ Poisson process. The rest of the setup is as before: there is a one server and the ith customer requires an amount of service s_i, where the s_i are independent and have a distribution G with mean $1/\mu$.

Let X_n be the number of customers in the queue when the nth customer enters service. To be precise, when $X_0 = x$, the chain starts with x people waiting in line and customer 0 just beginning her service. To understand the definition the following picture is useful:

To begin to define our Markov chain X_n, let

$$a_k = \int_0^\infty e^{-\lambda t} \frac{(\lambda t)^k}{k!} \, dG(t)$$

be the probability that k customers arrive during a service time.

$$\sum_k k a_k = \int_0^\infty \lambda t \, dG(t) = \lambda E s_i = \lambda/\mu$$

To construct the chain now let $\zeta_1, \zeta_2, \ldots$ be i.i.d. with $P(\zeta_i = k) = a_k$. We think of ζ_i as the number of customers to arrive during the ith service time. If $X_n > 0$ then

$$X_{n+1} = X_n + \zeta_n - 1.$$

In words ζ_n customers are added to the queue and then it shrinks by one when the nth customer departs see Fig. 3.1. If $X_n = 0$ and $\zeta_n > 0$, then the same logic applies but if $X_n = 0$ and $\zeta_n = 0$ then $X_{n+1} = 0$. To fix this we write

$$X_{n+1} = (X_n + \zeta_n - 1)^+ \tag{3.6}$$

where $z^+ = \max\{z, 0\}$ is the positive part.

From this formula we see that the transition probability is

$$
\begin{array}{c c c c c c c}
 & \mathbf{0} & \mathbf{1} & \mathbf{2} & \mathbf{3} & \mathbf{4} & \mathbf{5} \ldots \\
\mathbf{0} & a_0 + a_1 & a_2 & a_3 & a_4 & a_5 & a_6 \\
\mathbf{1} & a_0 & a_1 & a_2 & a_3 & a_4 & a_5 \\
\mathbf{2} & 0 & a_0 & a_1 & a_2 & a_3 & a_4 \\
\mathbf{3} & 0 & 0 & a_0 & a_1 & a_2 & a_3 \\
\mathbf{4} & 0 & 0 & 0 & a_0 & a_1 & a_2
\end{array}
$$

To explain the value of $p(0,0)$ we note that if $X_n = 0$ and $\zeta_n = 0$ or 1 then $X_{n+1} = 0$. In all other cases if $X_{n+1} = X_n + \zeta_n - 1$, so, for example, $p(2,4) = P(\zeta_n = 3)$.

Our result for the GI/G/1 queue implies $1/E_0T_0 = \pi(0) = 1 - \lambda/\mu$ so we have proved the first part of

Theorem 3.7.   *If $\lambda < \mu$, then X_n is positive recurrent and $E_0T_0 = \mu/(\mu - \lambda)$.*
    If $\lambda = \mu$, then X_n is null recurrent.
    If $\lambda > \mu$, then X_n is transient.

Proof. Turning to the second conclusion, note that setting $\lambda = \mu$ in the first result gives $E_0T_0 = \infty$. To begin the proof of recurrence note that $E\zeta_i = 0$ so if $x > 0$ then $E_xX_1 = x$. Let $\tau(M) = \min\{n : X_n = 0 \text{ or } X_n \geq M\}$. Iterating we have

$$
x = E_x(X_{\tau(M)\wedge n}) \geq MP_x(X_{\tau(M)} \geq M, \tau(M) \leq n)
$$

since $X_n \geq 0$. So letting $n \to \infty$ gives

$$
P_x(X_{\tau(M)} \geq M) \leq x/M
$$

Letting $M \to \infty$ gives $P_x(T_0 < \infty) = 1$. To prove that the chain is transient when $\lambda > \mu$ note that if we consider the customers that arrive during a person's service time her children then we have a supercritical branching process.   □

Busy Periods We learned in Theorem 3.7 that if $\lambda < \mu$ then an $M/G/1$ queue will repeatedly return to the empty state. Thus the server experiences alternating busy periods with duration B_n and idle periods with duration I_n. The lack of memory property implies that I_n has an exponential distribution with rate λ. Combining this observation with our result for alternating renewal processes we see that the limiting fraction of time the server is idle is

$$
\frac{1/\lambda}{1/\lambda + EB_n} = \pi(0)
$$

by (3.5). Rearranging, and using (3.5) we have

$$
EB_n = \frac{1}{\lambda}\left(\frac{1}{\pi(0)} - 1\right) = \frac{1}{\lambda}\left(\frac{\mu}{\mu - \lambda} - 1\right)
$$

which simplifies to

$$EB_n = \frac{1}{\mu - \lambda} \tag{3.7}$$

PASTA These initials stand for "Poisson arrivals see time averages." To be precise, if $\pi(n)$ is the limiting fraction of time that there are n individuals in the queue and α_n is the limiting fraction of arriving customers that see a queue of size n, then

Theorem 3.8. $\alpha_n = \pi(n)$.

Why is this true? Suppose the process has been running since time $-\infty$ and hence the system is in equilibrium. If we condition on there being arrival at time t, then the times of the previous arrivals are a Poisson process with rate λ. Thus knowing that there is an arrival at time t does not affect the distribution of what happened before time t.   □

For a simple example to show this is not always true, consider a GI/G/1 queue in which service times always take one unit of time while interarrival times are uniform on $[2, 3]$. Since server alternates between being busy for time 1 and idle for an amount to time with mean 2.5, the limiting fraction of time the server is busy is $1/(1 + 2.5) = 2/7$. However, an arriving customer never sees a busy server.

Theorem 3.9 (Pollaczek–Khintchine Formula). *The long run average waiting time in queue*

$$W_Q = \frac{\lambda E(s_i^2/2)}{1 - \lambda E s_i} \tag{3.8}$$

Proof. We define the workload in the system at time t, Z_t, to be the sum of the remaining service times of all customers in the system, and define the long run average workload to be

$$Z = \lim_{t \to \infty} \frac{1}{t} \int_0^t Z_s \, ds$$

As in the proof of Little's formula we will derive our result by computing the rate at which revenue is earned in two ways. This time we suppose that each customer in the queue or in service pays at a rate of \$$y$ when his remaining *service* time is y; i.e., we do not count the remaining waiting time in the queue. The customer's payment is equal to her current contribution to the workload so by the reasoning for Little's formula:

$$Z = \lambda Y$$

Since a customer with service time s_i pays s_i during the q_i units of time spent waiting in the queue and at rate $s_i - x$ after x units of time in service

$$Y = E(s_i q_i) + E\left(\int_0^{s_i} s_i - x \, dx\right)$$

Now a customer's waiting time in the queue can be determined by looking at the arrival process and at the service times of previous customers, so it is independent of her service time, i.e., $E(s_i q_i) = Es_i \cdot W_Q$ and we have

$$Y = (Es_i)W_Q + E(s_i^2/2)$$

PASTA implies that arriving customers see the long run average behavior so the workload they see $Z = W_Q$, so we have

$$W_Q = \lambda(Es_i)W_Q + \lambda E(s_i^2/2)$$

Solving for W_Q gives the desired formula.   □

Using formula (3.3), and Theorem 3.6, we can now compute

$$W = W_Q + Es_i \qquad L = \lambda W$$

Example 3.7. We see a number of applications of the equations from this section to Markovian queues in Chap. 4. Customers arrive at a help desk at rate $1/6$ per minute, i.e., the mean time between arrivals is six minutes. Suppose that each service takes a time with mean 5 and standard deviation 7.

(a) In the long run what is the fraction of time, $\pi(0)$, that the server is idle?
$\lambda = 1/6$, $Es_i = 5 = 1/\mu$, so by (3.5) $\pi(0) = 1 - (1/6)/(1/5) = 1/6$.
(b) What is the average waiting W time for a customer (including their service time)? $Es_i^2 = 5^2 + 49 = 74$, so (3.8) implies

$$W_Q = \frac{\lambda Es_i^2/2}{1 - \lambda Es_i} = \frac{(1/6)\cdot 74/2}{1/6} = 37$$

and $W = W_Q + Es_i = 42$.
(c) What is the average queue length (counting the customer in service)?
By Little's formula, $L = \lambda W = 42/6 = 7$.

3.3   Age and Residual Life*

Let $t_1, t_2, \ldots$ be i.i.d. interarrival times, let $T_n = t_1 + \cdots + t_n$ be the time of the nth renewal, and let $N(t) = \max\{n : T_n \le t\}$ be the number of renewals by time t. Let

$$A(t) = t - T_{N(t)} \qquad \text{and} \qquad Z(t) = T_{N(t)+1} - t$$

$A(t)$ gives the age of the item in use at time t, while $Z(t)$ gives its residual lifetime (Fig. 3.2).

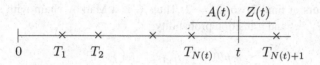

Fig. 3.2 Age and residual life

To explain the interest in $Z(t)$ note that the interrarival times after $T_{N(t)+1}$ will be independent of $Z(t)$ and i.i.d. with distribution F, so if we can show that $Z(t)$ converges in distribution, then the renewal process after time t will converge to an equilibrium.

3.3.1 Discrete Case

The situation in which all the interarrival times are positive integers is very simple but also important because it includes visits of a Markov chain to a fixed state, Example 3.1, are a special case. Let

$$V_m = \begin{cases} 1 & \text{if } m \in \{T_0, T_1, T_2, \ldots\} \\ 0 & \text{otherwise} \end{cases}$$

$V_m = 1$ if a renewal occurs at time m, i.e., if T_n visits m. Let $A_n = \min\{n - m : m \le n, V_m = 1\}$ be the age and let $Z_n = \min\{m - n : m \ge n, V_m = 1\}$ be the residual life. An example should help clarify the definitions:

n	0	1	2	3	4	5	6	7	8	9	10	11	12	13
V_n	1	0	0	0	1	0	0	1	1	0	0	0	0	1
A_n	0	1	2	3	0	1	2	0	0	1	2	3	4	0
Z_n	0	3	2	1	0	2	1	0	0	4	3	2	1	0

As we can see from the concrete example, values taken in an excursion away from 0 are $j, j - 1, \ldots 1$ in the residual life chain and $1, 2, \ldots j$ in the age chain so we will have

$$\lim_{n \to \infty} \frac{1}{n} \sum_{m=1}^{n} P(A_m = i) = \lim_{n \to \infty} \frac{1}{n} \sum_{m=1}^{n} P(Z_m = i)$$

From this we see that it is enough to study one of the two chains. We choose Z_n since it is somewhat simpler. It is clear that if $Z_n = i > 0$ then $Z_{n+1} = i - 1$.

When $Z_n = 0$, a renewal has just occurred. If the time to the next renewal is k, then $Z_{n+1} = k - 1$. To check this note that $Z_4 = 0$ and the time to the next renewal is 3 (it occurs at time 7) so $Z_5 = 2$. Thus Z_n is a Markov chain with state space $S = \{0, 1, 2, \ldots\}$ and transition probability

$$p(0,j) = f_{j+1} \quad \text{for } j \geq 0$$

$$p(i, i-1) = 1 \quad \text{for } i \geq 1$$

$$p(i,j) = 0 \quad \text{otherwise}$$

In this chain 0 is always recurrent. If there are infinitely many values of k with $f_k > 0$, then it is irreducible. If not and K is the largest value of k with $f_k > 0$, then $\{0, 1, \ldots K - 1\}$ is a closed irreducible set.

To define a stationary measure we will use the cycle trick, Theorem 1.24, with $x = 0$. Starting from 0 the chain will visit a site i at most once before it returns to 0, and this will happen if and only if the first jump is to a state $\geq i$, i.e., $t_1 > i$. Thus the stationary measure is

$$\mu(i) = P(t_1 > i)$$

Using (A.20) we see that

$$\sum_{i=0}^{\infty} \mu(i) = Et_1$$

so the chain is positive recurrent if and only if $Et_1 < \infty$. In this case

$$\pi(i) = P(t_1 > i)/Et_1 \tag{3.9}$$

$I_0 \supset J_0 = \{k : f_k > 0\}$ so if the greatest common divisor of J_0 is 1 then 0 is aperiodic. To argue the converse note that I_0 consists of all finite sums of elements in J_0 so g.c.d. I_0 = g.c.d. J_0. Using the Markov chain convergence theorem now gives:

Theorem 3.10. *Suppose $Et_1 < \infty$ and the greatest common divisor of $\{k : f_k > 0\}$ is 1 then*

$$\lim_{n \to \infty} P(Z_n = i) = \frac{P(t_1 > i)}{Et_1}$$

In particular $P(Z_n = 0) \to 1/Et_1$.

Example 3.8 (Visits to Go). In Monopoly one rolls two dice and then moves that number of squares. As in Example 1.23 we will ignore *Go to Jail, Chance,* and other squares that make the chain complicated. The average number of spaces moved in one roll is $Et_1 = 7$ so in the long run we land exactly on Go in 1/7 of the trips around the board. Using Theorem 3.10 we can calculate the limiting distribution of the amount we overshoot Go.

0	1	2	3	4	5	6	7	8	9	10	11
$\frac{1}{7}$	$\frac{1}{7}$	$\frac{35}{252}$	$\frac{33}{252}$	$\frac{30}{252}$	$\frac{26}{252}$	$\frac{21}{252}$	$\frac{15}{252}$	$\frac{10}{252}$	$\frac{6}{252}$	$\frac{3}{252}$	$\frac{1}{252}$

3.3.2 General Case

With the discrete case taken care of, we will proceed to the general case, which will be studied using renewal reward processes.

Theorem 3.11. *As* $t \to \infty$

$$\frac{1}{t} \int_0^t 1_{\{A_s > x, Z_s > y\}} \, ds \to \frac{1}{Et_1} \int_{x+y}^\infty P(t_i > z) \, dz$$

Proof. Let $I_{x,y}(s) = 1$ if $A_s > x$ and $Z_s > y$. It is easy to see that

$$\int_{T_{i-1}}^{T_i} I_c(s) \, ds = (t_i - (x + y))^+$$

To check this we consider two cases. Ignoring the contribution from the last incomplete cycle $[T_{N(t)}, t]$, we have

$$\int_0^t I_{x,y}(s) \, ds \approx \sum_{i=1}^{N(t)} (t_i - (x + y))^+$$

The right-hand side is a renewal reward process with so it follows from Theorem 3.3 that the limit is $E(t_1 - (x + y))^+ / Et_1$. Applying (A.22) to $X = (t_1 - (x + y))^+$ we have

$$E(t_1 - (x + y))^+ = \int_0^\infty P((t_1 - (x + y))^+ > u) \, du = \int_{x+y}^\infty P(t_i > z) \, dz$$

which proves the desired result. □

Setting $x = 0$ and then $y = 0$ we see that the limiting distribution for the age and residual life have the density function given by

$$g(z) = \frac{P(t_i > z)}{Et_i} \tag{3.10}$$

which looks the same as the result in discrete time. Differentiating twice we see that if t_i has density function f_T then the limiting joint density of (A_t, Z_t) is

$$f_T(a+z)/Et_1 \tag{3.11}$$

Example 3.9 (Exponential). In this case the limiting density given in (3.11) is

$$\frac{\lambda e^{-\lambda(a+z)}}{1/\lambda} = \lambda e^{-\lambda a} \cdot \lambda e^{-\lambda z}$$

So in the limit the age and residual life are independent exponential.

Example 3.10 (Uniform on (0,b)). Plugging into (3.11) gives for $a, z > 0, a+z < b$:

$$\frac{1/b}{b/2} = \frac{2}{b^2}$$

i.e., the limiting joint density is uniform on the triangle of possible values. The marginal densities given in (3.10) are

$$\frac{(b-x)/b}{b/2} = \frac{2}{b} \cdot \left(1 - \frac{x}{b}\right)$$

In words, the limiting density is a linear function that starts at $2/b$ at 0 and hits 0 at $c = b$.

Inspection Paradox Multiplying the formula in (3.10) by z, integrating from 0 to ∞ and using (A.21):

$$\int_0^\infty zP(t_i > z)\, dz = Et_i^2/2.$$

That is, the limit distribution in Theorem 3.11 has expected value

$$Et_i^2/2Et_i \tag{3.12}$$

Let $L(t) = A(t) + Z(t)$ be the lifetime of the item in use at time t. Using the last result, we see that

$$\frac{1}{t}\int_0^t L(s)\, ds \to \frac{E(t_1^2)}{Et_1} > Et_i$$

since var $(t_i) = Et_i^2 - (Et_i)^2 > 0$. This is a paradox because the average of the lifetimes of the first n items:

$$\frac{t_1 + \cdots + t_n}{n} \to Et_i$$

and hence

$$\frac{t_1 + \cdots + t_{N(t)}}{N(t)} \to Et_i$$

There is a simple explanation for this "paradox": taking the average age of the item in use up to time s is biased since items that last for time u are counted u times. That is,

$$\frac{1}{t} \int_0^t L(s) \, ds \approx \frac{N(t)}{t} \cdot \frac{1}{N(t)} \sum_{i=1}^{N(t)} t_i \cdot t_i \to \frac{1}{Et_1} \cdot Et_1^2$$

3.4   Chapter Summary

This chapter shows the power of the law of large numbers to give simple derivations of useful results. The work horse of the chapter is the result for renewal reward processes. If the times between renewals and the rewards earned in these periods (t_i, r_i) are an i.i.d. sequence, then the limiting rate at which rewards are earned is Er_i/Et_i. Taking $r_i = 1$ this reduces to the law of large numbers (Theorem 3.1) for the renewal process

$$N(t)/t \to 1/Et_i$$

If the $t_i = s_i + u_i$ with the (s_i, u_i) are i.i.d. representing the times in states 1 and 2, then taking $r_i = s_i$ we get see that the limiting fraction of time in state 1 is

$$Es_i/(Es_i + Eu_i)$$

our result (Theorem 3.4) for alternating renewal processes. Other applications of renewal reward processes gave us results for the limiting behavior of the age and residual life in Sect. 3.3.

A second theme here was the simple minded scheme of computing costs two different ways to prove quantities were equal. For the $GI/G/1$ queue, this allowed us to show that if

the average interarrival time $Et_i = 1/\lambda$,
the average service time $Es_i = 1/\mu$,
the average waiting time in the queue is L,
the long run rate at which customers enter the system is λ_a,
the average waiting time in the system is W,
and the fraction of time the queue is empty is $\pi(0)$

then we have

$$L = \lambda_a W \qquad \pi(0) = 1 - \frac{\lambda_a}{\mu}$$

In the $M/G/1$ case, the expected duration of busy periods and the average waiting time in the queue satisfy:

$$\pi(0) = \frac{1/\lambda}{1/\lambda + EB} \qquad W_Q = \frac{E(s_i^2/2)}{1 - \lambda/\mu}$$

The first formula is a simple consequence of our result for alternating renewal. The more sophisticated second formula uses "Poisson Arrivals See Time Averages" along with cost equation reasoning.

3.5   Exercises

3.1. The weather in a certain locale consists of alternating wet and dry spells. Suppose that the number of days in each rainy spell is a Poisson distribution with mean 2, and that a dry spell follows a geometric distribution with mean 7. Assume that the successive durations of rainy and dry spells are independent. What is the long-run fraction of time that it rains?

3.2. Monica works on a temporary basis. The mean length of each job she gets is 11 months. If the amount of time she spends between jobs is exponential with mean 3 months, then in the long run what fraction of the time does she spend working?

3.3. Thousands of people are going to a Grateful dead concert in Pauley Pavillion at UCLA. They park their 10 foot cars on several of the long streets near the arena. There are no lines to tell the drivers where to park, so they park at random locations, and end up leaving spacings between the cars that are independent and uniform on $(0, 10)$. In the long run, what fraction of the street is covered with cars?

3.4. The times between the arrivals of customers at a taxi stand are independent and have a distribution F with mean μ_F. Assume an unlimited supply of cabs, such as might occur at an airport. Suppose that each customer pays a random fare with distribution G and mean μ_G. Let $W(t)$ be the total fares paid up to time t. Find $\lim_{t \to \infty} E\, W(t)/t$.

3.5. In front of terminal C at the Chicago airport is an area where hotel shuttle vans park. Customers arrive at times of a Poisson process with rate 10 per hour looking for transportation to the Hilton hotel nearby. When seven people are in the van it leaves for the 36-minute round trip to the hotel. Customers who arrive while the van is gone go to some other hotel instead. (a) What fraction of the customers actually go to the Hilton? (b) What is the average amount of time that a person who actually goes to the Hilton ends up waiting in the van?

3.6. Three children take turns shooting a ball at a basket. They each shoot until they miss and then it is next child's turn. Suppose that child i makes a basket with probability p_i and that successive trials are independent. (a) Determine the proportion of time in the long run that each child shoots. (b) Find the answer when $p_1 = 2/3, p_2 = 3/4, p_3 = 4/5$.

3.7. A policeman cruises (on average) approximately ten minutes before stopping a car for speeding. 90 % of the cars stopped are given speeding tickets with an $80 fine. It takes the policeman an average of five minutes to write such a ticket. The other 10 % of the stops are for more serious offenses, leading to an average fine of $300. These more serious charges take an average of 30 minutes to process. In the long run, at what rate (in dollars per minute) does he assign fines.

3.8. *Counter processes.* Suppose that arrivals at a counter come at times of a Poisson process with rate λ. An arriving particle that finds the counter free gets registered and then locks the counter for an amount of time τ. Particles that arrive while the counter is locked have no effect. Find the limiting probability the counter is locked at time t. (b) Compute the limiting fraction of particles that get registered.

3.9. A cocaine dealer is standing on a street corner. Customers arrive at times of a Poisson process with rate λ. The customer and the dealer then disappear from the street for an amount of time with distribution G while the transaction is completed. Customers that arrive during this time go away never to return. (a) At what rate does the dealer make sales? (b) What fraction of customers are lost?

3.10. One of the difficulties about probability is realizing when two different looking problems are the same, in this case dealing cocaine and fighting fires. In Problem 2.19, calls to a fire station arrive according to a Poisson process with rate 0.5 per hour, and the time required to respond to a call, return to the station, and get ready to respond to the next call is uniformly distributed between 1/2 and 1 hour. If a new call comes before the Dryden fire department is ready to respond, the Ithaca fire department is asked to respond. What fraction of calls must be handled by the Ithaca fire department

3.11. A young doctor is working at night in an emergency room. Emergencies come in at times of a Poisson process with rate 0.5 per hour. The doctor can only get to sleep when it has been 36 minutes (.6 hours) since the last emergency. For example, if there is an emergency at 1:00 and a second one at 1:17 then she will not be able to get to sleep until at least 1:53, and it will be even later if there is another emergency before that time.

(a) Compute the long-run fraction of time she spends sleeping, by formulating a renewal reward process in which the reward in the ith interval is the amount of time she gets to sleep in that interval.
(b) The doctor alternates between sleeping for an amount of time s_i and being awake for an amount of time u_i. Use the result from (a) to compute Eu_i.
(c) Solve problem (b) by noting that the doctor trying to sleep is the same as chicken crossing the road in Exercise 2.23.

3.12. A worker has a number of machines to repair. Each time a repair is completed a new one is begun. Each repair independently takes an exponential amount of time with rate μ to complete. However, independent of this, mistakes occur according to a Poisson process with rate λ. Whenever a mistake occurs, the item is ruined and work is started on a new item. In the long run how often are jobs completed?

3.13. In the Duke versus Wake Forest football game, possessions alternate between Duke who has the ball for an average of 3.5 minutes and Wake Forest who has the ball for an average of 2.5 minutes. (a) In the long run what fraction of time does Duke have the ball? (b) Suppose that on each possession Duke scores a touchdown with probability 2/7 and a field goal with probability 1/3, while Wake Forest scores a touchdown with probability 1/7 and a field goal with probability 1/2. On the average how many points will each team score per hour?

3.14. Random Investment. An investor has $100,000. If the current interest rate is $i\%$ (compounded continuously so that the grow per year is $\exp(i/100)$), he invests his money in a i year CD, takes the profits, and then reinvests the $100,000. Suppose that the kth investment leads to an interest rate X_k which is uniform on $\{1, 2, 3, 4, 5\}$. In the long run how much money does he make per year.

3.15. Consider the setup of Example 3.3 but now suppose that the car's lifetime $h(t) = \lambda e^{-\lambda t}$. Show that for any A and B the optimal time $T = \infty$. Can you give a simple verbal explanation?

3.16. A machine tool wears over time and may fail. The failure time measured in months has density $f_T(t) = 2t/900$ for $0 \le t \le 30$ and 0 otherwise. If the tool fails, it must be replaced immediately at a cost of $1200. If it is replaced prior to failure, the cost is only $300. Consider a replacement policy in which the tool is replaced after c months or when it fails. What is the value of c that minimizes cost per unit time.

3.17. People arrive at a college admissions office at rate 1 per minute. When k people have arrive a tour starts. Student tour guides are paid $20 for each tour they conduct. The college estimates that it loses ten cents in good will for each minute a person waits. What tour group size minimizes the average cost person?

3.18. A scientist has a machine for measuring ozone in the atmosphere that is located in the mountains just north of Los Angeles. At times of a Poisson process with rate 1, storms or animals disturb the equipment so that it can no longer collect data. The scientist comes every L units of time to check the equipment. If the equipment has been disturbed, then she can usually fix it quickly so we will assume the repairs take 0 time. (a) What is the limiting fraction of time the machine is working? (b) Suppose that the data that is being collected is worth a dollars per unit time, while each inspection costs $c < a$. Find the best value of the inspection time L.

Age and Residual Life

3.19. Consider the discrete renewal process with $f_j = P(t_1 = j)$ and $F_i = P(t_1 > i)$.
(a) Show that the age chain has transition probability

$$q(j, j+1) = \frac{F_{j+1}}{F_j} \qquad q(j, 0) = 1 - \frac{F_{j+1}}{F_j} = \frac{f_{j+1}}{F_j} \quad \text{for } j \geq 0$$

(b) Show that if $Et_1 < \infty$, the stationary distribution $\pi(i) = P(t_1 > i)/Et_1$. (c) Let
$p(i,j)$ be the transition probability for the renewal chain. Verify that q is the dual
chain of p, i.e., the chain p run backwards. That is,

$$q(i,j) = \frac{\pi(j)p(j,i)}{\pi(i)}$$

3.20. Show that chain in Exercise 1.39 with transition probability is

	1	2	3	4
1	1/2	1/2	0	0
2	2/3	0	1/3	0
3	3/4	0	0	1/4
4	1	0	0	0

is a special case of the age chain. Use this observation and the previous exercise to
compute the stationary distribution.

3.21. The city of Ithaca, New York, allows for two-hour parking in all downtown
spaces. Methodical parking officials patrol the downtown area, passing the same
point every two hours. When an official encounters a car, he marks it with chalk. If
the car is still there two hours later, a ticket is written. Suppose that you park your
car for a random amount of time that is uniformly distributed on $(0, 4)$ hours. What
is the probability you will get a ticket?

3.22. Each time the frozen yogurt machine at the mall breaks down, it is replaced by
a new one of the same type. (a) What is the limiting age distribution for the machine
in use if the lifetime of a machine has a gamma$(2,\lambda)$ distribution, i.e., the sum of
two exponentials with mean $1/\lambda$. (b) Find the answer to (a) by thinking about a rate
one Poisson process in which arrivals are alternately colored red and blue.

3.23. While visiting Haifa, Sid Resnick discovered that people who wish to travel
from the port area up the mountain frequently take a shared taxi known as a sherut.
The capacity of each car is five people. Potential customers arrive according to a
Poisson process with rate λ. As soon as five people are in the car, it departs for The
Carmel, and another taxi moves up to accept passengers on. A local resident (who
has no need of a ride) wanders onto the scene. What is the distribution of the time
he has to wait to see a cab depart?

3.24. Suppose that the limiting age distribution in (3.10) is the same as the original
distribution. Conclude that $F(x) = 1 - e^{-\lambda x}$ for some $\lambda > 0$.

Chapter 4
Continuous Time Markov Chains

4.1 Definitions and Examples

In Chap. 1 we considered Markov chains X_n with a discrete time index $n = 0, 1, 2, \ldots$ In this chapter we will extend the notion to a continuous time parameter $t \geq 0$, a setting that is more convenient for some applications. In discrete time we formulated the Markov property as: for any possible values of $j, i, i_{n-1}, \ldots i_0$

$$P(X_{n+1} = j | X_n = i, X_{n-1} = i_{n-1}, \ldots, X_0 = i_0) = P(X_{n+1} = j | X_n = i)$$

In continuous time, it is technically difficult to define the conditional probability given all of the X_r for $r \leq s$, so we instead say that $X_t, t \geq 0$ is a Markov chain if for any $0 \leq s_0 < s_1 \cdots < s_n < s$ and possible states $i_0, \ldots, i_n, i, j$ we have

$$P(X_{t+s} = j | X_s = i, X_{s_n} = i_n, \ldots, X_{s_0} = i_0) = P(X_t = j | X_0 = i)$$

In words, given the present state, the rest of the past is irrelevant for predicting the future. Note that built into the definition is the fact that the probability of going from i at time s to j at time $s + t$ only depends on t the difference in the times.

Our first step is to construct a large collection of examples. In Example 4.6 we will see that this is almost the general case.

Example 4.1. Let $N(t)$, $t \geq 0$ be a Poisson process with rate λ and let Y_n be a discrete time Markov chain with transition probability $u(i,j)$. Then $X_t = Y_{N(t)}$ is a continuous time Markov chain. In words, X_t takes one jump according to $u(i,j)$ at each arrival of $N(t)$.

Why is this true? Intuitively, this follows from the lack of memory property of the exponential distribution. If $X_s = i$, then independent of what has happened in the past, the time to the next jump will be exponentially distributed with rate λ and will go to state j with probability $u(i,j)$. □

© Springer International Publishing Switzerland 2016
R. Durrett, *Essentials of Stochastic Processes*, Springer Texts in Statistics,
DOI 10.1007/978-3-319-45614-0_4

Discrete time Markov chains were described by giving their transition probabilities $p(i,j)$ = the probability of jumping from i to j in one step. In continuous time there is no first time $t > 0$, so we introduce for each $t > 0$ a **transition probability**

$$p_t(i,j) = P(X_t = j|X_0 = i)$$

To compute this for Example 4.1, we note that $N(t)$ has a Poisson number of jumps with mean λt, so

$$p_t(i,j) = \sum_{n=0}^{\infty} e^{-\lambda t} \frac{(\lambda t)^n}{n!} u^n(i,j)$$

where $u^n(i,j)$ is the nth power of the transition probability $u(i,j)$.

In continuous time, as in discrete time, the transition probability satisfies

Theorem 4.1 (Chapman–Kolmogorov Equation).

$$\sum_k p_s(i,k)p_t(k,j) = p_{s+t}(i,j)$$

Why is this true? In order for the chain to go from i to j in time $s + t$, it must be in some state k at time s, and the Markov property implies that the two parts of the journey are independent.

Proof. Breaking things down according to the state at time s, we have

$$P(X_{s+t} = j|X_0 = i) = \sum_k P(X_{s+t} = j, X_s = k|X_0 = i)$$

Using the definition of conditional probability and the Markov property, the above is

$$= \sum_k P(X_{s+t} = j|X_s = k, X_0 = i)P(X_s = k|X_0 = i) = \sum_k p_t(k,j)p_s(i,k) \quad \square$$

(4.1) shows that if we know the transition probability for $t < t_0$ for any $t_0 > 0$, we know it for all t. This observation and a large leap of faith (which we will justify later) suggest that the transition probabilities p_t can be determined from their derivatives at 0:

$$q(i,j) = \lim_{h \to 0} \frac{p_h(i,j)}{h} \qquad \text{for } j \neq i \tag{4.1}$$

If this limit exists (and it will in all the cases we consider) we will call $q(i,j)$ the **jump rate** from i to j. To explain this name we will compute the:

Jump Rates for Example 4.1 The probability of at least two jumps by time h is 1 minus the probability of 0 or 1 jumps

$$1 - \left(e^{-\lambda h} + \lambda h e^{-\lambda h}\right) = 1 - (1 + \lambda h)\left(1 - \lambda h + \frac{(\lambda h)^2}{2!} + \ldots\right)$$

$$= (\lambda h)^2/2! + \ldots = o(h)$$

That is, when we divide it by h it tends to 0 as $h \to 0$. Thus, if $j \neq i$,

$$\frac{p_h(i,j)}{h} \approx \lambda e^{-\lambda h} u(i,j) \to \lambda u(i,j)$$

as $h \to 0$. Comparing the last equation with the definition of the jump rate in (4.1) we see that $q(i,j) = \lambda u(i,j)$. In words we leave i at rate λ and go to j with probability $u(i,j)$. □

Example 4.1 is atypical. There we started with the Markov chain and then computed its rates. In most cases, it is much simpler to describe the system by writing down its transition rates $q(i,j)$ for $i \neq j$, which describe the rates at which jumps are made from i to j. The simplest possible example is

Example 4.2 (Poisson Process). Let $X(t)$ be the number of arrivals up to time t in a Poisson process with rate λ. Since arrivals occur at rate λ in the Poisson process the number of arrivals, $X(t)$, increases from n to $n + 1$ at rate λ, or in symbols

$$q(n, n + 1) = \lambda \quad \text{for all } n \geq 0$$

Example 4.3 (M/M/s Queue). Imagine a bank with s tellers that serve customers who queue in a single line if all of the servers are busy. We suppose that customers arrive at times of a Poisson process with rate λ, and that each service time is an independent exponential with rate μ. As in Example 4.2, $q(n, n+1) = \lambda$. To model the departures we let

$$q(n, n - 1) = \begin{cases} n\mu & 0 \leq n \leq s \\ s\mu & n \geq s \end{cases}$$

To explain this, we note that when there are $n \leq s$ individuals in the system then they are all being served and departures occur at rate $n\mu$. When $n > s$, all s servers are busy and departures occur at $s\mu$. The fact that the time of the next departure is exponential with the indicated rates follows from our discussion of exponential races.

Example 4.4 (Branching Process). In this system each individual dies at rate μ and gives birth to a new individual at rate λ so we have

$$q(n, n+1) = \lambda n \qquad q(n, n-1) = \mu n$$

A very special case called the **Yule process** occurs when $\mu = 0$.

Having seen several examples, it is natural to ask:
Given the rates, how do you construct the chain?

Let $\lambda_i = \sum_{j \neq i} q(i, j)$ be the rate at which X_t leaves i. If $\lambda_i = \infty$, then the process will want to leave i immediately, so we will always suppose that each state i has $\lambda_i < \infty$. If $\lambda_i = 0$, then X_t will never leave i. When $\lambda_i > 0$ we let

$$r(i, j) = q(i, j)/\lambda_i \qquad \text{for } j \neq i.$$

Here r, short for "routing matrix," is the probability the chain goes to j when it leaves i.

Informal Construction If X_t is in a state i with $\lambda_i = 0$, then X_t stays there forever and the construction is done. If $\lambda_i > 0$, X_t stays at i for an exponentially distributed amount of time with rate λ_i, then goes to state j with probability $r(i, j)$.

Formal Construction Suppose, for simplicity, that $\lambda_i > 0$ for all i. Let Y_n be a Markov chain with transition probability $r(i, j)$. The discrete time chain Y_n gives the road map that the continuous time process will follow. To determine how long the process should stay in each state let $\tau_0, \tau_1, \tau_2, \ldots$ be independent exponentials with rate 1.

At time 0 the process is in state Y_0 and should stay there for an amount of time that is exponential with rate $\lambda(Y_0)$, so we let the time the process stays in state Y_0 be $t_1 = \tau_0/\lambda(Y_0)$.

At time $T_1 = t_1$ the process jumps to Y_1, where it should stay for an exponential amount of time with rate $\lambda(Y_1)$, so we let the time the process stays in state Y_1 be $t_2 = \tau_1/\lambda(Y_1)$.

At time $T_2 = t_1 + t_2$ the process jumps to Y_2, where it should stay for an exponential amount of time with rate $\lambda(Y_2)$, so we let the time the process stays in state Y_2 be $t_3 = \tau_2/\lambda(Y_2)$.

Continuing in the obvious way, we can let the amount of time the process stays in Y_n be $t_{n+1} = \tau_n/\lambda(Y_n)$, so that the process jumps to Y_{n+1} at time

$$T_{n+1} = t_1 + \cdots + t_{n+1}$$

In symbols, if we let $T_0 = 0$, then for $n \geq 0$ we have

$$X(t) = Y_n \quad \text{for } T_n \leq t < T_{n+1} \tag{4.2}$$

Computer Simulation The construction described above gives a recipe for simulating a Markov chain. Generate independent standard exponentials τ_i, say, by looking at $\tau_i = -\ln U_i$ where U_i are uniform on $(0, 1)$. Using another sequence of random numbers, generate the transitions of Y_n, then define t_i, T_n, and X_t as above.

The good news about the formal construction above is that if $T_n \to \infty$ as $n \to \infty$, then we have succeeded in defining the process for all time and we are done. This will be the case in almost all the examples we consider. The bad news is that $\lim_{n\to\infty} T_n < \infty$ can happen. In most models, it is senseless to have the process make an infinite amount of jumps in a finite amount of time so we introduce a "cemetery state" Δ to the state space and complete the definition by letting $T_\infty = \lim_{n\to\infty} T_n$ and setting

$$X(t) = \Delta \quad \text{for all } t \geq T_\infty$$

To show that explosions can occur we consider

Example 4.5 (Pure Birth Processes with Power Law Rates). Suppose $q(i, i + 1) = \lambda i^p$ and all the other $q(i, j) = 0$. In this case the jump to $n + 1$ is made at time $T_n = t_1 + \cdots + t_n$, where t_n is exponential with rate n^p. $Et_m = 1/m^p$, so if $p > 1$

$$ET_n = \lambda \sum_{m=1}^{n} 1/m^p$$

This implies $ET_\infty = \sum_{m=1}^{\infty} 1/m^p < \infty$, so $T_\infty < \infty$ with probability one. When $p = 1$ which is the case for the Yule process

$$ET_n = (1/\beta) \sum_{m=1}^{n} 1/m \sim (\log n)/\beta$$

as $n \to \infty$. This is, by itself, not enough to establish that $T_n \to \infty$, but it is not hard to fill in the missing details.

Proof. $\text{var}(T_n) = \sum_{m=1}^{n} 1/m^2\beta^2 \leq C = \sum_{m=1}^{\infty} 1/m^2\beta^2$. Chebyshev's inequality implies

$$P(T_n \leq ET_n/2) \leq 4C/(ET_n)^2 \to 0$$

as $n \to \infty$. Since $n \to T_n$ is increasing, it follows that $T_n \to \infty$. $\square$

Our final example justifies the remark we made before Example 4.1.

Example 4.6 (Uniformization). Suppose that $\Lambda = \sup_i \lambda_i < \infty$ and let

$$u(i, j) = q(i, j)/\Lambda \qquad \text{for } j \neq i$$
$$u(i, i) = 1 - \lambda_i/\Lambda$$

In words, each site attempts jumps at rate Λ but stays put with probability $1 - \lambda_i/\Lambda$ so that the rate of leaving state i is λ_i. If we let Y_n be a Markov chain with transition probability $u(i, j)$ and $N(t)$ be a Poisson process with rate Λ, then $X_t = Y_{N(t)}$ has the desired transition rates. This construction is useful because Y_n is simpler to simulate that $X(t)$.

4.2   Computing the Transition Probability

In the last section we saw that given jump rates $q(i, j)$ we can construct a Markov chain that has these jump rates. This chain, of course, has a transition probability

$$p_t(i, j) = P(X_t = j | X_0 = i)$$

Our next question is: How do you compute the transition probability p_t from the jump rates q?

Our road to the answer starts by using the Chapman–Kolmogorov equations, Theorem 4.1, and then taking the $k = i$ term out of the sum.

$$p_{t+h}(i, j) - p_t(i, j) = \left(\sum_k p_h(i, k) p_t(k, j) \right) - p_t(i, j)$$

$$= \left(\sum_{k \neq i} p_h(i, k) p_t(k, j) \right) + [p_h(i, i) - 1] p_t(i, j) \qquad (4.3)$$

Our goal is to divide each side by h and let $h \to 0$ to compute

$$p'_t(i, j) = \lim_{h \to 0} \frac{p_{t+h}(i, j) - p_t(i, j)}{h}$$

By the definition of the jump rates

$$q(i, j) = \lim_{h \to 0} \frac{p_h(i, j)}{h} \quad \text{for } i \neq j$$

Ignoring the detail of interchanging the limit and the sum, which we will do throughout this chapter, we have

$$\lim_{h \to 0} \frac{1}{h} \sum_{k \neq i} p_h(i, k) p_t(k, j) = \sum_{k \neq i} q(i, k) p_t(k, j) \qquad (4.4)$$

For the other term we note that $1 - p_h(i, i) = \sum_{k \neq i} p_h(i, k)$, so

$$\lim_{h \to 0} \frac{p_h(i, i) - 1}{h} = -\lim_{h \to 0} \sum_{k \neq i} \frac{p_h(i, k)}{h} = -\sum_{k \neq i} q(i, k) = -\lambda_i$$

and we have

$$\lim_{h\to 0} \frac{p_h(i,i) - 1}{h} p_t(i,j) = -\lambda_i p_t(i,j) \tag{4.5}$$

Combining (4.4) and (4.5) with (4.3) and the definition of the derivative we have

$$p'_t(i,j) = \sum_{k\neq i} q(i,k)p_t(k,j) - \lambda_i p_t(i,j) \tag{4.6}$$

To neaten up the last expression we introduce a new matrix

$$Q(i,j) = \begin{cases} q(i,j) & \text{if } j \neq i \\ -\lambda_i & \text{if } j = i \end{cases}$$

For future computations note that the off-diagonal elements $q(i,j)$ with $i \neq j$ are nonnegative, while the diagonal entry is a negative number chosen to make the row sum equal to 0.

Using matrix notation we can write (4.6) simply as

$$p'_t = Qp_t \tag{4.7}$$

This is **Kolmogorov's backward equation**. If Q were a number instead of a matrix, the last equation would be easy to solve. We would set $p_t = e^{Qt}$ and check by differentiating that the equation held. Inspired by this observation, we define the matrix

$$e^{Qt} = \sum_{n=0}^{\infty} \frac{(Qt)^n}{n!} = \sum_{n=0}^{\infty} Q^n \cdot \frac{t^n}{n!} \tag{4.8}$$

and check by differentiating that

$$\frac{d}{dt}e^{Qt} = \sum_{n=1}^{\infty} Q^n \frac{t^{n-1}}{(n-1)!} = \sum_{n=1}^{\infty} Q \cdot \frac{Q^{n-1}t^{n-1}}{(n-1)!} = Qe^{Qt}$$

Kolmogorov's Forward Equation This time we split $[0, t+h]$ into $[0, t]$ and $[t, t+h]$ rather than into $[0, h]$ and $[h, t+h]$.

$$p_{t+h}(i,j) - p_t(i,j) = \left(\sum_k p_t(i,k)p_h(k,j) \right) - p_t(i,j)$$

$$= \left(\sum_{k\neq j} p_t(i,k)p_h(k,j) \right) + [p_h(j,j) - 1]p_t(i,j)$$

Computing as before we arrive at

$$p'_t(i,j) = \sum_{k \neq j} p_t(i,k)q(k,j) - p_t(i,j)\lambda_j \tag{4.9}$$

Introducing matrix notation again, we can write

$$p'_t = p_t Q \tag{4.10}$$

Comparing (4.10) with (4.7) we see that $p_t Q = Q p_t$ and that the two forms of Kolmogorov's differential equations correspond to writing the rate matrix on the left or the right. While we are on the subject of the choices, we should remember that in general for matrices $AB \neq BA$, so it is somewhat remarkable that $p_t Q = Q p_t$. The key to the fact that these matrices commute is that $p_t = e^{Qt}$ is made up of powers of Q:

$$Q \cdot e^{Qt} = \sum_{n=0}^{\infty} Q \cdot \frac{(Qt)^n}{n!} = \sum_{n=0}^{\infty} \frac{(Qt)^n}{n!} \cdot Q = e^{Qt} \cdot Q$$

To illustrate the use of Kolmogorov's equations we will now consider some examples. The simplest possible is

Example 4.7 (Poisson Process). Let $X(t)$ be the number of arrivals up to time t in a Poisson process with rate λ. In order to go from i arrivals at time s to j arrivals at time $t + s$ we must have $j \geq i$ and have exactly $j - i$ arrivals in t units of time, so

$$p_t(i,j) = e^{-\lambda t} \frac{(\lambda t)^{j-i}}{(j-i)!} \tag{4.11}$$

To check the differential equation, we have to first figure out what it is. Using (4.6) and plugging in our rates, we have

$$p'_t(i,j) = \lambda p_t(i+1,j) - \lambda p_t(i,j)$$

To check this we have to differentiate the formula in (4.11).
When $j > i$ we have that the derivative of (4.11) is

$$-\lambda e^{-\lambda t} \frac{(\lambda t)^{j-i}}{(j-i)!} + e^{-\lambda t} \frac{(\lambda t)^{j-i-1}}{(j-i-1)!} \lambda = -\lambda p_t(i,j) + \lambda p_t(i+1,j)$$

When $j = i$, $p_t(i,i) = e^{-\lambda t}$, so the derivative is

$$-\lambda e^{-\lambda t} = -\lambda p_t(i,i) = -\lambda p_t(i,i) + \lambda p_t(i+1,i)$$

since $p_t(i+1,i) = 0$.

The second simplest example is

Example 4.8 (Two State Chains). For concreteness, we can suppose that the state space is $\{1, 2\}$. In this case, there are only two flip rates $q(1, 2) = \lambda$ and $q(2, 1) = \mu$, so when we fill in the diagonal with minus the sum of the flip rates on that row we get

$$Q = \begin{pmatrix} -\lambda & \lambda \\ \mu & -\mu \end{pmatrix}$$

Writing out the backward equation in matrix form, (4.7), now we have

$$\begin{pmatrix} p_t'(1, 1) & p_t'(1, 2) \\ p_t'(2, 1) & p_t'(2, 2) \end{pmatrix} = \begin{pmatrix} -\lambda & \lambda \\ \mu & -\mu \end{pmatrix} \begin{pmatrix} p_t(1, 1) & p_t(1, 2) \\ p_t(2, 1) & p_t(2, 2) \end{pmatrix}$$

Since $p_t(i, 2) = 1 - p_t(i, 1)$ it is sufficient to compute $p_t(i, 1)$. Doing the first column of matrix multiplication on the right, we have

$$p_t'(1, 1) = -\lambda p_t(1, 1) + \lambda p_t(2, 1) = -\lambda(p_t(1, 1) - p_t(2, 1))$$

$$p_t'(2, 1) = \mu p_t(1, 1) - \mu p_t(2, 1) = \mu(p_t(1, 1) - p_t(2, 1)) \qquad (4.12)$$

Taking the difference of the two equations gives

$$[p_t(1, 1) - p_t(2, 1)]' = -(\lambda + \mu)[p_t(1, 1) - p_t(2, 1)]$$

Since $p_0(1, 1) = 1$ and $p_0(2, 1) = 0$ we have

$$p_t(1, 1) - p_t(2, 1) = e^{-(\lambda+\mu)t}$$

Using this in (4.12) and integrating

$$p_t(1, 1) = p_0(1, 1) + \frac{\lambda}{\mu + \lambda} e^{-(\mu+\lambda)s} \Big|_0^t = \frac{\mu}{\lambda + \mu} + \frac{\lambda}{\mu + \lambda} e^{-(\mu+\lambda)t}$$

$$p_t(2, 1) = p_0(2, 1) + \frac{\lambda}{\mu + \lambda} e^{-(\mu+\lambda)s} \Big|_0^t = \frac{\mu}{\mu + \lambda} - \frac{\mu}{\mu + \lambda} e^{-(\mu+\lambda)t}$$

As a check on the constants note that $p_0(1, 1) = 1$ and $p_0(2, 1) = 0$. To prepare for the developments in the next section note that the probability of being in state 1 converges exponentially fast to the equilibrium value $\mu/(\mu + \lambda)$.

Example 4.9 (Jukes–Cantor Model). In this chain the states are the four nucleotides A, C, G, T. Jumps, which correspond to nucleotide substitutions, occur according at rate $q(x, y) = \mu$ if $x \neq y$. To find the transition probability, we modify the chain so that jumps occur at rate 4μ and the new state is uniform on $\{A, C, G, T\}$. Since the first jump brings the chain to equilibrium:

$$p_t(x, y) = (1/4)(1 - e^{-4\mu t}) \quad y \neq x$$
$$p_t(x, x) = e^{-4\mu t} + (1/4)(1 - e^{-4\mu t})$$

To illustrate the use of this formula, let $v = 3\mu t$ be the expected number of changes per site. If we observe a frequency f of nucleotide differences, then

$$f = (3/4)(1 - e^{-4v/3})$$

Rearranging we have the following estimate of v:

$$v = -\frac{3}{4} \ln(1 - 4f/3)$$

Example 4.10 (Kimura Two Parameter Model). The nucleotides A and G are purines while C's and T's are pyrimidines. Kimura's model takes into account that mutations that do not change the type of base (called transitions) happen at a different rate than those that do (called transversions) so the transition matrix becomes

	A	G	C	T
A	$-(\alpha + 2\beta)$	α	β	β
G	α	$-(\alpha + 2\beta)$	β	β
C	β	β	$-(\alpha + 2\beta)$	α
T	β	β	α	$-(\alpha + 2\beta)$

Let X_t be the Markov chain with this transition probability. If we let $Y_t = B$ when $X_t \in \{A, G\}$ and $Y_t = D$ when $X_t \in \{C, T\}$, then Y_t is a two state Markov chain that jumps at rate 2β so by the result in Example 4.8, the transition probability $\bar{p}_t$ for Y_t is given by:

$$p_t(B, B) = \frac{1}{2}(1 + e^{-4\beta t})$$

$$p_t(B, D) = \frac{1}{2}(1 - e^{-4\beta t})$$

Let $p_1(t) = p_t(A, A)$, $p_2(t) = p_t(A, G)$, and $p_3(t) = p_t(A, C) = p_t(A, T)$.

$$p_1(t) + p_2(t) = p_t(B, B) = \frac{1}{2}(1 + e^{-4\beta t}) \tag{4.13}$$

$$p_3(t) = p_t(B, D)/2 = \frac{1}{4}(1 - e^{-4\beta t}) \tag{4.14}$$

so it remains to compute $p_1(t) - p_2(t)$. Using the forward equation (4.10) we have

$$p_1'(t) = -(\alpha + 2\beta)p_1(t) + \alpha p_2(t) + 2\beta \cdot 2p_3(t)$$
$$p_2'(t) = \alpha p_1(t) - (\alpha + 2\beta)p_2(t) + 2\beta \cdot 2p_3(t)$$

so we have

$$(p_1(t) - p_2(t))' = -(2\alpha + 2\beta)(p_1(t) - p_2(t)),$$

and it follows that $p_1(t) - p_2(t) = e^{-(2\alpha+2\beta)t}$. Using this with (4.13) we have

$$p_1(t) = \frac{1}{4}(1 + e^{-4\beta t}) + \frac{1}{2}e^{-(2\alpha+2\beta)t}$$

$$p_2(t) = \frac{1}{4}(1 + e^{-4\beta t}) - \frac{1}{2}e^{-(2\alpha+2\beta)t}.$$

As in the previous example if we see a fraction f_1 of transitions and f_2 of transitions per site we can use these formulas to estimate αt and βt.

There are a number of other substitution matrices that include more details of the process but the solutions become increasingly complicated. For a concrete example we list the one of Hasekawa, Kishino, and Yano (1985)

	A	G	C	T
A	$-\lambda_A$	$\kappa\pi_G$	π_C	π_G
G	$\kappa\pi_A$	$-\lambda_G$	π_C	π_G
C	π_A	π_C	$-\lambda_C$	$\kappa\pi_G$
T	π_A	π_C	$\kappa\pi_C$	$-\lambda_T$

It allows for unequal nucleotide frequencies, which can be estimated from the data. The model is reversible. κ is the ratio of transition to transversion rates. We have written λ_i on the diagonal to avoid writing the sums of the rows.

4.2.1 Branching Processes

Example 4.11 (Yule Process). In this process, which we call Y_t each particle splits into two at rate β, so $q(i, i+1) = \beta i$.

Theorem 4.2. *The transition probability of the Yule process is given by*

$$p_t(1, j) = e^{-\beta t}(1 - e^{-\beta t})^{j-1} \qquad \text{for } j \geq 1 \tag{4.15}$$

$$p_t(i, j) = \binom{j-1}{i-1}(e^{-\beta t})^i(1 - e^{-\beta t})^{j-i} \tag{4.16}$$

That is, $p_t(1,j)$ a geometric distribution with success probability $e^{-\beta t}$ and hence mean $e^{\beta t}$. To explain the mean we note that

$$\frac{d}{dt}EY_t = \beta EY_t \quad \text{implies} \quad E_1Y_t = e^{\beta t}.$$

From (4.15) we get the following nice consequence:

$$P(\exp(-\beta t)Y_t > x) = P(Y_t > xe^{\beta t}) = (1 - 1/e^{\beta t})^{xe^\beta} \to e^{-x}$$

In words $e^{-\beta t}Y_t$ converges to a mean one exponential. This is just the general fact that if X_μ has a geometric distribution with mean μ, then X_μ/μ converges to a mean one exponential as $\mu \to \infty$.

Proof. To check (4.15), we will use the forward equation (4.9) to conclude that if $j \geq 1$, then

$$p_t'(1,j) = -\beta j p_t(1,j) + \beta(j-1)p_t(1,j-1) \tag{4.17}$$

where $p_t(1,0) = 0$. The use of the forward equation here is dictated by the fact that we are deriving a formula for $p_t(1,j)$. To check the proposed formula for $j = 1$ we note that

$$p_t'(1,1) = -\beta e^{-\beta t} = -\beta p_t(1,1)$$

Things are not so simple for $j > 1$:

$$p_t'(1,j) = -\beta e^{-\beta t}(1 - e^{-\beta t})^{j-1}$$
$$+ e^{-\beta t}(j-1)(1 - e^{-\beta t})^{j-2}(\beta e^{-\beta t})$$

Recopying the first term on the right and using $\beta e^{-\beta t} = -(1 - e^{-\beta t})\beta + \beta$ in the second, we can rewrite the right-hand side of the above as

$$-\beta e^{-\beta t}(1 - e^{-\beta t})^{j-1} - e^{-\beta t}(j-1)(1 - e^{-\beta t})^{j-1}\beta$$
$$+ e^{-\beta t}(1 - e^{-\beta t})^{j-2}(j-1)\beta$$

Adding the first two terms then comparing with (4.17) shows that the above is

$$= -\beta j p_t(1,j) + \beta(j-1)p_t(1,j-1)$$

Having worked to find $p_t(1,j)$, it is fortunately easy to find $p_t(i,j)$. The chain starting with i individuals is the sum of i independent copies of the chain starting from 1 individual. To compute $p_t(i,j)$ now, we note that if $N_1, \ldots N_i$ have the distribution given in (4.15) and $n_1 + \cdots + n_i = j$, then

$$P(N_1 = n_1, \ldots, N_i = n_i) = \prod_{k=1}^{i} e^{-\beta t}(1 - e^{-\beta t})^{n_k-1} = (e^{-\beta t})^i (1 - e^{-\beta t})^{j-i}$$

To count the number of possible $(n_1, \ldots, n_i)$ with $n_k \geq 1$ and sum j, we think of putting j balls in a row. To divide the j balls into i groups of size $n_1, \ldots, n_i$, we will insert cards in the slots between the balls and let n_k be the number of balls in the kth group. Having made this transformation it is clear that the number of $(n_1, \ldots, n_i)$ is the number of ways of picking $i - 1$ of the $j - 1$ slots to put the cards or $\binom{j-1}{i-1}$. Multiplying this times the probability for each $(n_1, \ldots, n_i)$ gives the result. □

A Simple Proof of (4.15) Let $T_k = \min\{t : Y_t = k\}$. For $1 \leq m \leq k-1$, $T_{m+1} - T_m$ is exponential with rate βm. Let $X_1, X_2, \ldots X_{k-1}$ be independent exponential with rate β and let $V_1 < V_2 \ldots < V_k$ be these variables arranged in increasing order. If we let $V_0 = 0$, then $V_{m+1} - V_m$ is exponential with rate $\beta(k - m)$ due to the lack of memory property of the exponential and what we know about exponential races. Combining the last two observations:

$$T_k =_d \max\{X_1, \ldots X_{k-1}\}$$

From this it follows that $P(T_k \leq t) = (1 - e^{-\beta t})^{k-1}$ and

$$P(Y_t = k) = P(T_k \leq t < T_{k+1}) = P(T_k \leq t) - P(T_{k+1} \leq t) = e^{-\beta t}(1 - e^{-\beta t})^{k-1}$$

Example 4.12 (The General Case). Consider the process $Z(t)$ in which each individual gives birth at rate λ and dies at rate μ. That is, the transition rates are

$$q(i, i+1) = \lambda i, \qquad q(i, i-1) = \mu i,$$

and $q(i, j) = 0$ otherwise. As in the previous example it is enough to consider what happens when $Z(0) = 1$. Since each individual gives birth at rate λ and dies as rate μ

$$\frac{d}{dt}EZ(t) = (\lambda - \mu)EZ(t),$$

so $E_1 Z(t) = \exp((\lambda - \mu)t)$. Because of this we will restrict our attention to the case $\lambda > \mu$.

To compute the transition probability, we will find the generating function $F(x, t) = Ex^{Z_0(t)}$.

Lemma 4.3. $\partial F/\partial t = -(\lambda + \mu)F + \lambda F^2 + \mu = (1 - F)(\mu - \lambda F)$.

Proof. The second equality is just algebra. If h is small, then the probability of more than one event in $[0, h]$ is $O(h^2)$, the probability of a birth is $\approx \lambda h$, of a death is $\approx \mu h$. In the second case we have no particles, so the generating function of $Z_0(t + h)$ will be $\equiv 1$. In the first case we have two particles at time h who give rise to two independent copies of the branching process, so the generating function of $Z_0(t + h)$ will be $F(x, t)^2$. Combining these observations,

$$F(x, t + h) = \lambda h F(x, t)^2 + \mu h \cdot 1 + (1 - (\lambda + \mu)h)F(x, t) + O(h^2).$$

A little algebra converts this into

$$\frac{F(x, t + h) - F(x, t)}{h} = \lambda F(x, t)^2 + \mu - (\lambda + \mu)F(x, t) + O(h).$$

Letting $h \to 0$ gives the desired result.   □

Let $\rho = \lambda - \mu$. Our goal is to show

$$F(x, t) = \frac{\mu(x - 1) - e^{-\rho t}(\lambda x - \mu)}{\lambda(x - 1) - e^{-\rho t}(\lambda x - \mu)}. \tag{4.18}$$

Proof. To solve (4.3) we fix x and let $g(t) = F(x, t)$. Rearranging the differential equation in Lemma 4.3 gives

$$dt = \frac{dg}{(1 - g)(\mu - \lambda g)} = \frac{1}{\mu - \lambda}\left(\frac{dg}{1 - g} - \lambda\frac{dg}{\mu - \lambda g}\right).$$

Recalling $\rho = \lambda - \mu$ and integrating we have for some constant D

$$t + D = \frac{1}{\rho}\log(1 - g) - \frac{1}{\rho}\log(\mu - \lambda g) = -\frac{1}{\rho}\log\left(\frac{\mu - \lambda g}{1 - g}\right), \tag{4.19}$$

so we have

$$\mu - \lambda g = e^{-\rho t}e^{-D\rho}(1 - g) \tag{4.20}$$

Solving for g gives

$$g(t) = \frac{\mu - e^{-D\rho}e^{-\rho t}}{\lambda - e^{-D\rho}e^{-\rho t}}. \tag{4.21}$$

Taking $t = 0$ in (4.20) and using $g(0) = x$ we have $\mu - \lambda x = e^{-D\rho}(1 - x)$

$$e^{-D\rho} = \frac{\lambda x - \mu}{x - 1}.$$

Using this in (4.21) and rearranging gives (4.18).   □

It is remarkable that one can invert the generating function to find the underlying distribution. These formulas come from Section 8.6 in Bailey (1964).

Lemma 4.4. *If we let*

$$\alpha = \frac{\mu e^{\rho t} - \mu}{\lambda e^{\rho t} - \mu} \quad \text{and} \quad \beta = \frac{\lambda e^{\rho t} - \lambda}{\lambda e^{\rho t} - \mu}, \tag{4.22}$$

then Z_t has a generalized geometric distribution

$$p_0 = \alpha \qquad p_n = (1-\alpha)(1-\beta)\beta^{n-1} \quad \text{for } n \geq 1. \tag{4.23}$$

Proof. To check this claim, note that taking $x = 0$ in (4.18) and multiplying top and bottom by $-e^{\rho t}$ confirms the size of the atom at 0 and suggests that we write

$$F(x,t) = \frac{(\mu e^{\rho t} - \mu) - x(\mu e^{\rho t} - \lambda)}{(\lambda e^{\rho t} - \mu) - x(\lambda e^{\rho t} - \lambda)}. \tag{4.24}$$

For comparison we note that the generating function of (4.23) is

$$\alpha + (1-\alpha)(1-\beta)\sum_{n=1}^{\infty} \beta^{n-1}x^n = \alpha + (1-\alpha)(1-\beta)\frac{x}{1-\beta x}$$

$$= \frac{\alpha + (1-\alpha-\beta)x}{1-\beta}.$$

Dividing the numerator and the denominator in (4.24) by $\lambda e^{\rho t} - \mu$ gives

$$F(x,t) = \frac{\alpha - x(\mu e^{\rho t} - \lambda)/(\lambda e^{\rho t} - \mu)}{1 - \beta x},$$

Using (4.22) we see

$$1 - \alpha - \beta = \frac{\lambda e^{\rho t} - \mu - \mu e^{\rho t} + \mu - \lambda e^{\rho t} + \lambda}{\lambda e^{\rho t} - \mu} = -\frac{\mu e^{\rho t} - \lambda}{\lambda e^{\rho t} - \mu}.$$

which completes the proof.   □

Having worked to prove an ugly formula for the generating function and then to invert it, we can prove a nice result.

Theorem 4.5. *Suppose $\lambda > \mu$. As $t \to \infty$, $e^{-\rho t}Z_0(t)$ converges in distribution to W with $P(W = 0) = \mu/\lambda$ and*

$$P(W > x | W > 0) = \exp(-x\rho/\lambda) \tag{4.25}$$

In words, $(W|W > 0)$ is exponential with rate ρ/λ.

Proof. From (4.23), the atom at 0

$$\alpha(t) = \frac{\mu - \mu e^{-\rho t}}{\lambda - \mu e^{-\rho t}} \to \frac{\lambda}{\mu}.$$

The geometric distribution $(1 - \beta)\beta^{n-1}$ has mean

$$\frac{1}{1 - \beta} = \frac{\lambda e^{\rho t} - \mu}{\lambda - \mu} \sim \frac{\lambda e^{\rho t}}{\rho},$$

where $x_t \sim y_t$ means $x_t / y_t \to 1$. Since a geometric distribution rescaled by its mean converges to a mean 1 exponential the desired result follows.   □

4.3   Limiting Behavior

Having worked hard to develop the convergence theory for discrete time chains, the results for the continuous time case follow easily. In fact the study of the limiting behavior of continuous time Markov chains is simpler than the theory for discrete time chains, since the randomness of the exponential holding times implies that we don't have to worry about aperiodicity. We begin by generalizing some of the previous definitions

The Markov chain X_t is **irreducible**, if for any two states i and j it is possible to get from i to j in a finite number of jumps. To be precise, there is a sequence of states $k_0 = i, k_1, \ldots k_n = j$ so that $q(k_{m-1}, k_m) > 0$ for $1 \le m \le n$.

Lemma 4.6. *If X_t is irreducible and $t > 0$, then $p_t(i, j) > 0$ for all i, j.*

Proof. Since $p_s(i, j) \ge \exp(-\lambda_j s) > 0$ and $p_{t+s}(i, j) \ge p_t(i, j)p_s(j, j)$ it suffices to show that this holds for small t. Since

$$\lim_{h \to 0} p_h(k_{m-1}, k_m)/h = q(k_{m-1}, k_m) > 0$$

it follows that if h is small enough we have $p_h(k_{m-1}, k_m) > 0$ for $1 \le m \le n$ and hence $p_{nh}(i, j) > 0$.   □

In discrete time a stationary distribution is a solution of $\pi p = \pi$. Since there is no first $t > 0$, in continuous time we need the stronger notion: π is said to be a **stationary distribution** if $\pi p_t = \pi$ for all $t > 0$. The last condition is difficult to check since it involves all of the p_t, and as we have seen in the previous section, the p_t are not easy to compute. The next result solves these problems by giving a test for stationarity in terms of the basic data used to describe the chain, the matrix of transition rates

$$Q(i,j) = \begin{cases} q(i,j) & j \neq i \\ -\lambda_i & j = i \end{cases}$$

where $\lambda_i = \sum_{j \neq i} q(i,j)$ is the total rate of transitions out of i.

Lemma 4.7. π is a stationary distribution if and only if $\pi Q = 0$.

Why is this true? Filling in the definition of Q and rearranging, the condition $\pi Q = 0$ becomes

$$\sum_{k \neq j} \pi(k)q(k,j) = \pi(j)\lambda_j$$

If we think of $\pi(k)$ as the amount of sand at k, the right-hand side represents the rate at which sand leaves j, while the left gives the rate at which sand arrives at j. Thus, π will be a stationary distribution if for each j the flow of sand into j is equal to the flow out of j. $\qquad\square$

More Details If $\pi p_t = \pi$, then

$$0 = \frac{d}{dt}\pi p_t = \sum_i \pi(i)p_t'(i,j) = \sum_i \pi(i) \sum_k p_t(i,k)Q(k,j)$$

$$= \sum_k \sum_i \pi(i)p_t(i,k)Q(k,j) = \sum_k \pi(k)Q(k,j)$$

Conversely if $\pi Q = 0$

$$\frac{d}{dt}\left(\sum_i \pi(i)p_t(i,j)\right) = \sum_i \pi(i)p_t'(i,j) = \sum_i \pi(i) \sum_k Q(i,k)p_t(k,j)$$

$$= \sum_k \sum_i \pi(i)Q(i,k)p_t(k,j) = 0$$

Since the derivative is 0, πp_t is constant and must always be equal to π its value at 0. $\qquad\square$

Lemma 4.6 implies that for any $h > 0$, p_h is irreducible and aperiodic, so by Theorem 1.19

$$\lim_{n \to \infty} p_{nh}(i,j) = \pi(j).$$

From this it is intuitively clear that

Theorem 4.8. *If a continuous time Markov chain X_t is irreducible and has a stationary distribution π, then*

$$\lim_{t \to \infty} p_t(i,j) = \pi(j).$$

Proof. When $nh \le t \le (n+1)h$, $p_t(i,j) \ge e^{-\lambda_j h} p_{nh}(i,j)$. From this it follows that $\liminf_{t \to \infty} p_t(i,j) \ge \pi(j)$. If A is a finite set of states with $j \notin A$, then

$$\limsup_{t \to \infty} p_t(i,j) \le 1 - \sum_{k \in A} \liminf_{t \to \infty} p_t(i,j) \le 1 - \sum_{k \in A} \pi(k)$$

Letting $A \uparrow S - \{j\}$ the right-hand side decreases to $\pi(j)$ gives the desired result. □

We will now consider some examples.

Example 4.13 (L.A. Weather Chain). There are three states: $1 =$ sunny, $2 =$ smoggy, and $3 =$ rainy. The weather stays sunny for an exponentially distributed number of days with mean 3, then becomes smoggy. It stays smoggy for an exponentially distributed number of days with mean 4, then rain comes. The rain lasts for an exponentially distributed number of days with mean 1, then sunshine returns. Remembering that for an exponential the rate is 1 over the mean, the verbal description translates into the following Q-matrix:

$$
\begin{array}{cccc}
 & 1 & 2 & 3 \\
1 & -1/3 & 1/3 & 0 \\
2 & 0 & -1/4 & 1/4 \\
3 & 1 & 0 & -1 \\
\end{array}
$$

The relation $\pi Q = 0$ leads to three equations:

$$
\begin{aligned}
-\tfrac{1}{3}\pi_1 \qquad +\pi_3 &= 0 \\
\tfrac{1}{3}\pi_1 - \tfrac{1}{4}\pi_2 \qquad &= 0 \\
\tfrac{1}{4}\pi_2 - \pi_3 &= 0
\end{aligned}
$$

Adding the three equations gives $0 = 0$ so we delete the third equation and add $\pi_1 + \pi_2 + \pi_3 = 1$ to get an equation that can be written in matrix form as

$$(\pi_1 \ \pi_2 \ \pi_3) A = (0\ 0\ 1) \quad \text{where} \quad A = \begin{pmatrix} -1/3 & 1/3 & 1 \\ 0 & -1/4 & 1 \\ 1 & 0 & 1 \end{pmatrix}$$

This is similar to our recipe in discrete time. To find the stationary distribution of a k state chain, form A by taking the first $k-1$ columns of Q, add a column of 1's and then

$$(\pi_1 \ \pi_2 \ \pi_3) = (0\ 0\ 1) A^{-1}$$

i.e., the last row of A^{-1}. In this case we have

$$\pi(1) = 3/8, \qquad \pi(2) = 4/8, \qquad \pi(3) = 1/8$$

To check our answer, note that the weather cycles between sunny, smoggy, and rainy spending independent exponentially distributed amounts of time with means 3, 4, and 1, so the limiting fraction of time spent in each state is just the mean time spent in that state over the mean cycle time, 8.

Example 4.14 (Duke Basketball). To "simulate" a basketball game we use a four state Markov chain with four states

$$0 = \text{Duke on offense} \quad 2 = \text{UNC on offense}$$
$$1 = \text{Duke scores} \quad 3 = \text{UNC scores}$$

Here the states 1 and 3 include the time between the basket is made and the other time brings the ball up the court and across the center line.

The Duke offense keeps the ball for an exponential amount of time with mean 1/4 minute, ending with a score with probability 0.7, and turning the ball over to UNC with probability 0.3. After a score UNC needs an average of eight seconds (2/15 minute) to take the ball down the court and then they are on offense.

The UNC offense keeps the ball for an exponential amount of time with mean 1/3 minute, ending with a score with probability 2/3, and turning the ball over to Duke with probability 1/3. After a score Duke needs an average of six seconds (1/10 minute) to take the ball down the court and then they are on offense.

The transition rate matrix with rates per minute is

$$
\begin{array}{c|cccc}
 & 0 & 1 & 2 & 3 \\
\hline
0 & -4 & 2.8 & 1.2 & 0 \\
1 & 0 & -7.5 & 7.5 & 0 \\
2 & 1 & 0 & -3 & 2 \\
3 & 10 & 0 & 0 & -10 \\
\end{array}
$$

To find the stationary distribution we want to solve

$$
\begin{pmatrix} \pi_0 & \pi_1 & \pi_2 & \pi_3 \end{pmatrix}
\begin{pmatrix}
-4 & 2.8 & 1.2 & 1 \\
0 & -7.5 & 7.5 & 1 \\
1 & 0 & -3 & 1 \\
10 & 0 & 0 & 1
\end{pmatrix}
= \begin{pmatrix} 0 & 0 & 0 & 1 \end{pmatrix}
$$

The answer can be found by reading the fourth row of the inverse of the matrix:

$$
\pi_0 = \frac{75}{223} \quad \pi_1 = \frac{28}{223} \quad \pi_2 = \frac{100}{223} \quad \pi_3 = \frac{20}{223} \tag{4.26}
$$

The stationary distribution gives the fraction of time spent in the various states. To translate this into information about the score in the basketball game, we note that the average time in state 1 is 2/15 and the average time in state 3 is 1/10, so the number of baskets per minute for the two teams is

$$\frac{28/223}{2/15} = \frac{210}{223} = 0.9417 \qquad \frac{20/223}{1/10} = \frac{200}{223} = 0.8968 \tag{4.27}$$

Multiplying by 2 points per basket and 40 minutes per game yields 75.34 and 71.75 points per game, respectively.

For another derivation of the stationary distribution, consider the embedded jump chain which we write in general as

$$
\begin{array}{c|cccc}
 & 0 & 1 & 2 & 3 \\
\hline
0 & 0 & p & 1-p & 0 \\
1 & 0 & 0 & 1 & 0 \\
2 & 1-q & 0 & 0 & q \\
3 & 1 & 0 & 0 & 0 \\
\end{array}
$$

Since we already have a stationary distribution called π, let μ be the stationary distribution of the embedded jump chain. Between two visits to 0 there is exactly one visit to 2, so $\mu_0 = \mu_2 = c$. There is only one way into states 1 and 3 so $\mu_1 = cp$ and $\mu - 3 = cq$. From this we see that the stationary distribution is

$$\mu_0 = \mu_2 = \frac{1}{2+p+q} \qquad \mu_1 = \frac{p}{2+p+q} \qquad \mu_3 = \frac{q}{2+p+q}$$

In our special case $p = 0.7$ and $q = 2/3$ so

$$2+p+q = \frac{60+21+20}{30} = \frac{101}{30}.$$

and it follows that

$$\mu_0 = \mu_2 = \frac{30}{101} \qquad \mu_1 = \frac{21}{101} \qquad \mu_3 = \frac{20}{101}$$

To convert the stationary distribution μ_i for the embedded chain to the one in continuous chain we multiply μ_i by the mean holding times $1/\lambda_i$ and then renormalize so the sum is 1:

$$\pi_i = \frac{\mu_i/\lambda_i}{\sum_k \mu_k/\lambda_k}$$

To carry this out we note that

$$101 \sum_k \mu_k/\lambda_k = \frac{30}{4} + \frac{21}{7.5} + \frac{30}{3} + \frac{20}{10}$$

$$= \frac{15}{2} + \frac{14}{5} + 10 + 2 = \frac{223}{10}$$

so we have

$$\pi_0 = 7.5 \cdot \frac{10}{223} \quad \pi_1 = \frac{14}{5} \cdot \frac{10}{223} \quad \pi_2 = 10 \cdot \frac{10}{223} \quad \pi_3 = 2 \cdot \frac{10}{223}$$

which agrees with (4.26).

4.3.1 Detailed Balance Condition

Generalizing from discrete time we can formulate this condition as:

$$\pi(k)q(k,j) = \pi(j)q(j,k) \quad \text{for all } j \neq k \tag{4.28}$$

The reason for interest in this concept is

Theorem 4.9. *If (4.28) holds, then π is a stationary distribution.*

Why is this true? The detailed balance condition implies that the flows of sand between each pair of sites are balanced, which then implies that the net amount of sand flowing into each vertex is 0, i.e., $\pi Q = 0$.

Proof. Summing (4.28) over all $k \neq j$ and recalling the definition of λ_j gives

$$\sum_{k \neq j} \pi(k)q(k,j) = \pi(j) \sum_{k \neq j} q(j,k) = \pi(j)\lambda_j$$

Rearranging we have

$$(\pi Q)_j = \sum_{k \neq j} \pi(k)q(k,j) - \pi(j)\lambda_j = 0 \qquad \square$$

As in discrete time, (4.28) is much easier to check than $\pi Q = 0$ but does not always hold. In Example 4.13

$$\pi(2)q(2,1) = 0 < \pi(1)q(1,2)$$

As in discrete time, detailed balance holds for $\qquad \square$

Example 4.15 (Birth and Death Chains). Suppose that $S = \{0,1,\ldots,N\}$ with $N \leq \infty$ and

$$q(n,n+1) = \lambda_n \qquad \text{for } n < N$$
$$q(n,n-1) = \mu_n \qquad \text{for } n > 0$$

Here λ_n represents the birth rate when there are n individuals in the system, and μ_n denotes the death rate in that case.

If we suppose that all the λ_n and μ_n listed above are positive, then the birth and death chain is irreducible, and we can divide to write the detailed balance condition as

$$\pi(n) = \frac{\lambda_{n-1}}{\mu_n} \pi(n-1) \tag{4.29}$$

Using this again we have $\pi(n-1) = (\lambda_{n-2}/\mu_{n-1})\pi(n-2)$ and it follows that

$$\pi(n) = \frac{\lambda_{n-1}}{\mu_n} \cdot \frac{\lambda_{n-2}}{\mu_{n-1}} \cdot \pi(n-2)$$

Repeating the last reasoning leads to

$$\pi(n) = \frac{\lambda_{n-1} \cdot \lambda_{n-2} \cdots \lambda_0}{\mu_n \cdot \mu_{n-1} \cdots \mu_1} \pi(0) \tag{4.30}$$

To check this formula and help remember it, note that (i) there are n terms in the numerator and in the denominator, and (ii) if the state space was $\{0, 1, \ldots, n\}$, then $\mu_0 = 0$ and $\lambda_n = 0$, so these terms cannot appear in the formula.

To illustrate the use of (4.30) we will consider several concrete examples.

Example 4.16 (Two State Chains). Suppose that the state space is $\{1, 2\}$, $q(1, 2) = \lambda$, and $q(2, 1) = \mu$, where both rates are positive. The equations $\pi Q = 0$ can be written as

$$(\pi_1 \ \pi_2) \begin{pmatrix} -\lambda & \lambda \\ \mu & -\mu \end{pmatrix} = (0 \ 0)$$

The first equation says $-\lambda \pi_1 + \mu \pi_2 = 0$. Taking into account that we must have $\pi_1 + \pi_2 = 1$, it follows that

$$\pi_1 = \frac{\mu}{\lambda + \mu} \quad \text{and} \quad \pi_2 = \frac{\lambda}{\lambda + \mu}$$

Example 4.17 (M/M/∞ Queue). In this case $q(n, n+1) = \lambda$ and $q(n, n-1) = n\mu$ so

$$\pi(n) = \pi(0) \frac{(\lambda/\mu)^n}{n!}$$

If we take $\pi(0) = e^{-\lambda/\mu}$, then this becomes the Poisson distribution with mean λ/μ.

Example 4.18 (Two Barbers). Suppose that a shop has two barbers that can each cut hair at rate 3 people per hour customers arrive at times of a rate 2 Poisson process, but will leave if there are two people getting their haircut and two waiting. Find the stationary distribution for the number of customers in the shop.

The transition rate matrix is

	0	1	2	3	4
0	-2	2	0	0	0
1	3	-5	2	0	0
2	0	6	-8	2	0
3	0	0	6	-8	2
4	0	0	0	6	-6

The detailed balance conditions say

$$2\pi(0) = 3\pi(1), \quad 2\pi(1) = 6\pi(2), \quad 2\pi(2) = 6\pi(3), \quad 2\pi(3) = 6\pi(4)$$

Setting $\pi(0) = c$ and solving, we have

$$\pi(1) = \frac{2c}{3}, \quad \pi(2) = \frac{\pi(1)}{3} = \frac{2c}{9}, \quad \pi(3) = \frac{\pi(2)}{3} = \frac{2c}{27}, \quad \pi(4) = \frac{\pi(3)}{3} = \frac{2c}{81}.$$

The sum of the π's is $(81 + 54 + 18 + 6 + 2)c/81 = 161c/81$, so $c = 81/161$ and

$$\pi(0) = \frac{81}{161}, \quad \pi(1) = \frac{54}{161}, \quad \pi(2) = \frac{18}{161}, \quad \pi(3) = \frac{6}{161}, \quad \pi(4) = \frac{2}{161}.$$

Example 4.19 (Machine Repair Model). A factory has three machines in use and one repairman. Suppose each machine works for an exponential amount of time with mean 60 days between breakdowns, but each breakdown requires an exponential repair time with mean 4 days. What is the long run fraction of time all three machines are working?

Let X_t be the number of working machines. Since there is one repairman we have $q(i, i + 1) = 1/4$ for $i = 0, 1, 2$. On the other hand, the failure rate is proportional to the number of machines working, so $q(i, i - 1) = i/60$ for $i = 1, 2, 3$. Setting $\pi(0) = c$ and plugging into the recursion (4.29) gives

$$\pi(1) = \frac{\lambda_0}{\mu_1} \cdot \pi(0) = \frac{1/4}{1/60} \cdot c = 15c$$

$$\pi(2) = \frac{\lambda_1}{\mu_2} \cdot \pi(1) = \frac{1/4}{2/60} \cdot 15c = \frac{225c}{2}$$

$$\pi(3) = \frac{\lambda_2}{\mu_3} \cdot \pi(2) = \frac{1/4}{3/60} \cdot \frac{225c}{2} = \frac{1125c}{2}$$

Adding up the π's gives $(1125 + 225 + 30 + 2)c/2 = 1382c/2$ so $c = 2/1480$ and we have

$$\pi(3) = \frac{1125}{1382} \quad \pi(2) = \frac{225}{1382} \quad \pi(1) = \frac{30}{1382} \quad \pi(0) = \frac{2}{1382}$$

Thus in the long run all three machines are working $1125/1382 = 0.8140$ of the time.

4.4   Exit Distributions and Exit Times

In this section we generalize results from Sects. 1.9 and 1.10 to continuous time.

4.4.1   Exit Distribution

We will approach this first using the embedded jump chain with transition probability

$$r(i,j) = \frac{q(i,j)}{\lambda_i} \quad \text{where } \lambda_i = \sum_{j \neq i} q(i,j)$$

Let $V_k = \min \{t \geq 0 : X_t = k\}$ be the time of the first visit to x and let $T_k = \min\{t \geq 0 : X_t = k \text{ and } X_s \neq k \text{ for some } s < t \}$ be the time of the first return. The second definition is made complicated by the fact that is $X_0 = k$ then the chain stays at k for an amount of time that is exponential with rate λ_k.

Example 4.20. **Branching process** has jump rates $q(i, i+1) = \lambda i$ and $q(i, i-1) = \mu i$. 0 is an absorbing state but for $i \geq 1$ the i's cancel and we have

$$r(i, i+1) = \frac{\lambda}{\lambda + \mu} \quad r(i, i-1) = \frac{\mu}{\lambda + \mu}$$

Thus absorption at 0 is certain if $\lambda \leq \mu$ but if $\lambda > \mu$, then by (1.25) the probability of avoiding extinction is

$$P_1(T_0 = \infty) = 1 - \frac{\mu}{\lambda}$$

For another derivation let $\rho = P_1(T_0 < \infty)$. By considering what happens when the chain leaves 0 we have

$$\rho = \frac{\mu}{\lambda + \mu} \cdot 1 + \frac{\lambda}{\lambda + \mu} \cdot \rho^2$$

since starting from state 2 extinction occurs if and only if each individual's family line dies out. Rearranging gives

$$0 = \lambda\rho^2 - (\lambda + \mu)\rho + \mu = (\lambda\rho - \lambda)(\rho - \mu/\lambda)$$

The root we want is $\mu/\lambda < 1$.

As the last example shows, if we work with the embedded chain, then we can use the approach of Sect. 1.9 to compute exit distributions. We can also work directly with the Q-matrix. Let $V_D = \min\{t : X_t \in D\}$ and $T = V_A \wedge V_B$. Suppose that $C = S - (A \cup B)$ is finite and $P_x(T < \infty) > 0$ for all $x \in C$. If we have $h(a) = 1$ for $a \in A$, $h(b) = 0$ for $b \in B$, and

$$h(i) = \sum_{j \neq i} \frac{q(i,j)}{\lambda_i} h(j) \quad \text{for } i \in C$$

then by Theorem 1.28 $h(i) = P_i(V_A < V_B)$. Multiplying each side of the last equation by $\lambda_i = -Q(i,i)$ we have

$$-Q(i,i)h(i) = \sum_{j \neq i} Q(i,j)h(j)$$

which simplifies to

$$\sum_j Q(i,j)h(j) = 0 \quad \text{for } i \in C.$$

Let R be the part of the Q matrix with $i,j \in C$. If we let $v(i) = \sum_{j \in A} Q(i,j)$ for $i \in C$ and think of this as a column vector, then since $h(a) = 1$, $a \in A$ and $h(b) = 0$ for $b \in B$, the previous equation can be written as

$$-\sum_{j \in C} R(i,j)h(j) = v(i)$$

so we have

$$h(i) = (-R)^{-1}v \tag{4.31}$$

$(-R)^{-1}$ is the analogue of $(I - p)^{-1}$ in discrete time. Again $(-R)^{-1}(i,j)$ is the expected amount of time spent in state j when we start at i.

A concrete example will help explain the procedure.

Example 4.18 (Continued). A shop has two barbers that can cut hair at rate 3, people per hour customers arrive at times of a rate 2 Poisson process, but will leave if there are two people getting their haircut and two waiting. The state of the system to be the number of people in the shop. Find $P_i(V_0 < V_4)$ for $i = 1, 2, 3$.

The transition rate matrix is

	0	1	2	3	4
0	-2	2	0	0	0
1	3	-5	2	0	0
2	0	6	-8	2	0
3	0	0	6	-8	2
4	0	0	0	6	-6

The matrix R is the 3×3 matrix inside the box, while $Q(i, 0)$ is the column vector to its left. Using (4.31) now

$$h = (-R)^{-1} \begin{pmatrix} 3 \\ 0 \\ 0 \end{pmatrix} = \begin{pmatrix} 13/41 & 4/41 & 1/41 \\ 12/41 & 10/41 & 5/82 \\ 9/41 & 15/82 & 7/41 \end{pmatrix}^{-1} \begin{pmatrix} 3 \\ 0 \\ 0 \end{pmatrix} = \begin{pmatrix} 39/41 \\ 36/41 \\ 27/41 \end{pmatrix}$$

Example 4.21 (Office Hours). In Exercise 2.10. Ron, Sue, and Ted arrive at the beginning of a professor's office hours. The amount of time they will stay is exponentially distributed with means of 1, 1/2, and 1/3 hour. Part (b) of the question is to compute the probability each student is the last to leave.

If we describe the state of the Markov chain by the rates of the students that are left, with Ø to denote an empty office, then the Q-matrix is

	123	12	13	23	1	2	3	Ø
123	-6	3	2	1	0	0	0	0
12	0	-3	0	0	2	1	0	0
13	0	0	-4	0	3	0	1	0
23	0	0	0	-5	0	3	2	0
1	0	0	0	0	-1	0	0	1
2	0	0	0	0	0	-2	0	2
3	0	0	0	0	0	0	-3	3

(4.32)

The matrix R we want to solve our problem is

	123	12	13	23
123	-6	3	2	1
12	0	-3	0	0
13	0	0	-4	0
23	0	0	0	-5

while if we want to compute the probability each student is last we note that

$$Q(i,1) = \begin{pmatrix} 0 \\ 2 \\ 3 \\ 0 \end{pmatrix} \qquad Q(i,2) = \begin{pmatrix} 0 \\ 1 \\ 0 \\ 3 \end{pmatrix} \qquad Q(i,3) = \begin{pmatrix} 0 \\ 0 \\ 1 \\ 2 \end{pmatrix}$$

Inverting the matrix gives

$$(-R)^{-1} = \begin{pmatrix} 1/6 & 1/6 & 1/12 & 1/30 \\ 0 & 1/3 & 0 & 0 \\ 0 & 0 & 1/4 & 0 \\ 0 & 0 & 0 & 1/5 \end{pmatrix}$$

The last three rows are obvious. Starting from these three states the first jump takes us to a state with only one student left, so the diagonal elements are the mean holding times. The top left entry in the inverse is the 1/6 hour until the first student leaves. The other three on the first row are

$$\frac{1}{2} \cdot \frac{1}{3} \qquad \frac{1}{3} \cdot \frac{1}{4} \qquad \frac{1}{6} \cdot \frac{1}{5}$$

which are the probabilities we visit the state times the expected amount of time we spend there. Using (4.31) that the probabilities ρ_i that the three students are last are

$$\rho_1 = 2 \cdot \frac{1}{6} + 3 \cdot \frac{1}{12} = \frac{35}{60} \qquad \rho_2 = \frac{1}{6} + 3 \cdot \frac{1}{12} = \frac{35}{60} \qquad \rho_3 = \frac{1}{12} + 2 \cdot \frac{1}{30} = \frac{35}{60}.$$

4.4.2 Exit Times

To develop the analogue of (4.31) suppose that $C = S - A$ is finite and $P_i(V_A < \infty) > 0$ for each $i \in C$. Under these assumptions if $g(a) = 0$ for $a \in A$, and for $i \in C = S - A$

$$g(i) = \frac{1}{\lambda_i} + \sum_{j \neq i} \frac{q(i,j)}{\lambda_i} g(j)$$

then by Theorem 1.29, $g(i) = E_i V_A$. Multiplying each side by $\lambda_i = -Q(i,i)$ we have

$$-Q(i,i)g(i) = 1 + \sum_{j \neq i} Q(i,j)g(j)$$

which simplifies to

$$\sum_j Q(i,j)g(j) = -1 \quad \text{for } i \notin A.$$

Writing $\mathbf{1}$ for a vector of 1's the solution, writing R for the part of Q with $i,j, \in C$, and noting $g(a) = 0$ for $a \in A$.

$$g = (-R)^{-1}\mathbf{1} \tag{4.33}$$

This holds because $(-R)^{-1}(i,j)$ is the expected time spent at j when we start at i, so when we multiply by $\mathbf{1}$ we sum over all j and get the exit time.

Example 4.18 (Continued). If we want to compute $E_x T_0$, then $A = \{0\}$, so the matrix R is

$$
\begin{array}{ccccc}
 & 1 & 2 & 3 & 4 \\
1 & -5 & 2 & 0 & 0 \\
2 & 6 & -8 & 2 & 0 \\
3 & 0 & 6 & -8 & 2 \\
4 & 0 & 0 & 6 & -6 \\
\end{array}
$$

which has

$$
-R^{-1} = \begin{pmatrix}
1/3 & 1/9 & 1/27 & 1/81 \\
1/3 & 5/18 & 5/54 & 5/16 \\
1/3 & 5/18 & 7/27 & 7/81 \\
1/3 & 5/18 & 7/27 & 41/162
\end{pmatrix}
$$

Multiplying by $\mathbf{1}$ so we have

$$g(i) = (40/81, 119/162, 155/162, /91/81)^T$$

where T denotes transpose, i.e., g is a column vector. Note that in the answer if $i > j$, then $(-R)^{-1}(i,j) = (-R)^{-1}(j,j)$ since must hit j on our way from i to 0.

The last calculation shows $E_1 T_0 = 40/81$. We can compute this from the stationary distribution in Example 4.18 if we note that the time spent at zero (which is exponential with mean 1/2) followed by the time to return from 1 to 0 (which has mean $E_1 T_0$) is an alternating renewal process

$$\frac{81}{161} = \pi(0) = \frac{1/2}{1/2 + E_1 T_0}.$$

This gives $1 + 2E_1 T_0 = 161/81$, and it follows that $E_1 T_0 = 40/81$.

Example 4.21 (Office Hours Continued). Part (c) of Problem 2.10 is to compute the expected time until all students are gone. Letting R be the matrix in (4.32) with the last column deleted, the first row of $-R^{-1}$ is

$$1/6 \quad 1/6 \quad 1/12 \quad 1/30 \quad 7/12 \quad 2/15 \quad 1/20$$

The sum is 63/60, or one hour and three minutes. We have already explained the first four entries in the row. Similarly the last three entries in the row are

$$\frac{35}{60} \cdot 1 \qquad \frac{16}{60} \cdot \frac{1}{2} \qquad \frac{9}{60} \cdot \frac{1}{3}$$

where is the probability we visit the state times the amount of time we spend there.

Example 4.22. We now take a different approach to analyzing the Duke Basketball chain, Example 4.14. The transition rate matrix with rates per minute is

$$
\begin{array}{c|cccc}
 & \mathbf{0} & \mathbf{1} & \mathbf{2} & \mathbf{3} \\
\hline
\mathbf{0} & -4 & 2.8 & 1.2 & 0 \\
\mathbf{1} & 0 & -7.5 & 7.5 & 0 \\
\mathbf{2} & 1 & 0 & -3 & 2 \\
\mathbf{3} & 10 & 0 & 0 & -10
\end{array}
$$

Our first step is to compute $g(i) = E_i(V_1)$ for $i = 0, 2, 3$. Removing the row and column for 1:

$$
R = \begin{array}{c|ccc}
 & \mathbf{0} & \mathbf{2} & \mathbf{3} \\
\hline
\mathbf{0} & -4 & 1.2 & 0 \\
\mathbf{2} & 1 & -3 & 2 \\
\mathbf{3} & 10 & 0 & -10
\end{array}
$$

so using (4.33)

$$
g = -R^{-1}\mathbf{1} = \begin{pmatrix} 5/15 & 1/7 & 1/35 \\ 5/14 & 10/21 & 2/21 \\ 5/14 & 1/7 & 9/70 \end{pmatrix} \mathbf{1} = \begin{pmatrix} 37/70 \\ 13/14 \\ 22/35 \end{pmatrix}
$$

Starting from 1 we stay in the state for an exponential time with mean 2/15 and then go to 2 so

$$
E_1 T_1 = \frac{13}{14} + \frac{2}{15} = \frac{223}{210}
$$

and it follows that Duke scores 210/223 baskets per minute as we calculated in Example 4.14.

To calculate the rate at which UNC scores points we will calculate the average number of times they score between two Duke baskets. To do this we start by computing $h(i) = P_i(V_3 < V_1)$ for $i = 0, 2$. To do this we further reduce the matrix to

$$
\begin{array}{cc}
 & \mathbf{0} \quad \mathbf{2} \\
R = \mathbf{0} & -4 \;\; 1.2 \\
\mathbf{2} & 1 \;\; -3
\end{array}
$$

and note that $Q(i, 3) = (0, 2)^T$. Using (4.31) we have

$$
h = (-R)^{-1} \begin{pmatrix} 0 \\ 2 \end{pmatrix} = \begin{pmatrix} 5/18 & 1/9 \\ 5/54 & 10/27 \end{pmatrix} \begin{pmatrix} 0 \\ 2 \end{pmatrix} = \begin{pmatrix} 2/9 \\ 20/27 \end{pmatrix}
$$

If we let Z denote the number of baskets that UNC scores between two Duke baskets, then

$$
P(Z = 0) = P_2(V_1 < V_3) = 7/27,
$$

because after Duke scores (state 1) the chain always jumps to state 2.

$$
P(Z = 1) = P_2(V_3 < V_1)P_0(V_1 < V_3)
$$

since after UNC scores (state 3) Duke gets the ball back (state 0). Continuing with this reasoning we have that for $k \geq 1$

$$
P(Z \geq k) = P_2(V_3 < V_1)P_0(V_3 < V_1)^{k-1}
$$

From this it follows that

$$
EZ = \sum_{k=1}^{\infty} P(Z \geq k) = \frac{P_2(V_3 < V_1)}{1 - P_0(V_3 < V_1)} = \frac{20/27}{7/9} = \frac{20}{21}
$$

Thus UNC scores baskets at a rate equal to 20/21 times Duke does or 200/223 per minute.

4.5 Markovian Queues

In this section we will take a systematic look at the basic models of queueing theory that have Poisson arrivals and exponential service times. The arguments in Sect. 3.2 explain why we can be happy assuming that the arrival process is Poisson. The assumption of exponential services times is hard to justify, but here, it is a necessary evil. The lack of memory property of the exponential is needed for the queue length to be a continuous time Markov chain. We begin with the simplest examples.

4.5.1 Single Server Queues

Example 4.23 (M/M/1 Queue). In this system customers arrive to a single server facility at the times of a Poisson process with rate λ, and each requires an independent amount of service that has an exponential distribution with rate μ. From the description it should be clear that the transition rates are

$$q(n, n+1) = \lambda \qquad \text{if } n \geq 0$$
$$q(n, n-1) = \mu \qquad \text{if } n \geq 1$$

so we have a birth and death chain with birth rates $\lambda_n = \lambda$ and death rates $\mu_n = \mu$. Plugging into our formula for the stationary distribution, (4.30), we have

$$\pi(n) = \frac{\lambda_{n-1} \cdots \lambda_0}{\mu_n \cdots \mu_1} \cdot \pi(0) = \left(\frac{\lambda}{\mu}\right)^n \pi(0) \tag{4.34}$$

To have the sum 1, we pick $\pi(0) = 1 - (\lambda/\mu)$, and the resulting stationary distribution is the shifted geometric distribution

$$\pi(n) = \left(1 - \frac{\lambda}{\mu}\right)\left(\frac{\lambda}{\mu}\right)^n \qquad \text{for } n \geq 0 \tag{4.35}$$

It is comforting to note that this agrees with the idle time formula, (3.5), which says $\pi(0) = 1 - \lambda/\mu$.

Having determined the stationary distribution, we can now compute various quantities of interest concerning the queue. The number of people in the system in equilibrium has a shifted geometric distribution so

$$L = \frac{1}{1 - \lambda/\mu} - 1 = \frac{\mu}{\mu - \lambda} - \frac{\mu - \lambda}{\mu - \lambda} = \frac{\lambda}{\mu - \lambda}$$

All customers enter the system so using Little's formula

$$W = L/\lambda = \frac{1}{\mu - \lambda}$$

To get this result using Little's formula $L = \lambda W$ we note that by (3.7), the server's busy periods have mean

$$EB = \frac{1}{\mu - \lambda}.$$

Example 4.24 (M/M/1 Queue with a Finite Waiting Room). In this system customers arrive at the times of a Poisson process with rate λ. Customers enter service if there are $< N$ individuals in the system, but when there are N customers in the system, the new arrival leaves never to return. Once in the system, each customer requires an independent amount of service that has an exponential distribution with rate μ.

Lemma 4.10. *Let X_t be a Markov chain with a stationary distribution π that satisfies the detailed balance condition. Let Y_t be the chain constrained to stay in a subset A of the state space. That is, jumps which take the chain out of A are not allowed, but allowed jumps occur at the original rates. In symbols, $\bar{q}(x, y) = q(x, y)$ if $x, y \in A$ and 0 otherwise. Let $C = \sum_{y \in A} \pi(y)$. Then $v(x) = \pi(x)/C$ is a stationary distribution for Y_t.*

Proof. If $x, y \in A$, then detailed balance for X_t implies $\pi(x)q(x, y) = \pi(y)q(y, x)$. From this it follows that $v(x)\bar{q}(x, y) = v(y)\bar{q}(y, x)$ so v satisfies the detailed balance condition for Y_t. $\square$

It follows from Lemma 4.10 that

$$\pi(n) = \left(\frac{\lambda}{\mu}\right)^n / C \qquad \text{for } 1 \leq n \leq N$$

To compute the normalizing constant, we recall that if $\theta \neq 1$, then

$$\sum_{n=0}^{N} \theta^n = \frac{1 - \theta^{N+1}}{1 - \theta} \tag{4.36}$$

Suppose now that $\lambda \neq \mu$. Using (4.36), we see that

$$C = \frac{1 - (\lambda/\mu)^{N+1}}{1 - \lambda/\mu}$$

so the stationary distribution is given by

$$\pi(n) = \frac{1 - \lambda/\mu}{1 - (\lambda/\mu)^{N+1}} \left(\frac{\lambda}{\mu}\right)^n \qquad \text{for } 0 \leq n \leq N \tag{4.37}$$

The new formula is similar to the old one in (4.35) and when $\lambda < \mu$ reduces to it as $N \to \infty$. Of course, when the waiting room is finite, the state space is finite and we always have a stationary distribution, even when $\lambda > \mu$. The analysis above has been restricted to $\lambda \neq \mu$. However, it is easy to see that when $\lambda = \mu$ the stationary distribution is $\pi(n) = 1/(N + 1)$ for $0 \leq n \leq N$.

Example 4.25 (One Barber). Suppose now we have one barber who cuts hair at rate 3, customers arrive at rate 2 and there are two waiting chairs. In this case $N = 3$, $\lambda = 2$, and $\mu = 3$, so plugging into (4.37) and multiplying numerator and denominator by $3^4 = 81$, we have

$$\pi(0) = \frac{1 - 2/3}{1 - (2/3)^4} = \frac{81 - 54}{81 - 16} = 27/65$$

$$\pi(1) = \frac{2}{3}\pi(0) = 18/65$$

$$\pi(2) = \frac{2}{3}\pi(1) = 12/65$$

$$\pi(3) = \frac{2}{3}\pi(2) = 8/65$$

From the equation for the equilibrium we have that the average queue length

$$L = 1 \cdot \frac{18}{65} + 2 \cdot \frac{12}{65} + 3 \cdot \frac{8}{65} = \frac{66}{65}$$

Customers will only enter the system if there are < 3 people, so

$$\lambda_a = 2(1 - \pi(3)) = 114/65$$

and using the idle time formula (3.5)

$$\pi(0) = 1 - \frac{\lambda_a}{3} = 1 - \frac{114}{195} = \frac{81}{195} = \frac{27}{65}$$

Using Little's formula, Theorem 3.6, we see that the average waiting time for someone who enters the system is

$$W = \frac{L}{\lambda_a} = \frac{66/65}{114/65} = \frac{66}{114} = 0.579 \text{ hours}$$

From (3.7), it follows that the server's busy periods have mean

$$EB = \frac{1}{\lambda}\left(\frac{1}{\pi(0)} - 1\right) = \frac{1}{2}\left(\frac{65}{27} - 1\right) = \frac{19}{27}$$

To check this using the methods of the previous section we note that

$$R = \begin{matrix} & 1 & 2 & 3 \\ 1 & -5 & 2 & 0 \\ 2 & 3 & -5 & 2 \\ 3 & 0 & 3 & -3 \end{matrix}$$

so we have

$$(-R)^{-1}\mathbf{1} = \begin{pmatrix} 1/3 & 2/9 & 4/27 \\ 1/3 & 5/9 & 10/27 \\ 1/3 & 5/9 & 19/27 \end{pmatrix}\begin{pmatrix} 1 \\ 1 \\ 1 \end{pmatrix} = \begin{pmatrix} 19/27 \\ 34/27 \\ 43/27 \end{pmatrix}$$

Example 4.26 (M/M/1 Queue with Balking). Customers arrive at times of a Poisson process with rate λ but only join the queue with probability a_n if there are n customers in line. Thus, the birth rate $\lambda_n = \lambda a_n$ for $n \geq 0$, while the death rate is $\mu_n = \mu$ for $n \geq 1$. It is reasonable to assume that if the line is long the probability the customer will join the queue is small. The next result shows that this is always enough to prevent the queue length from growing out of control.

Theorem 4.11. *If $a_n \to 0$ as $n \to \infty$, then there is a stationary distribution.*

Proof. It follows from (4.29) that

$$\pi(n+1) = \frac{\lambda_n}{\mu_{n+1}} \cdot \pi(n) = \frac{a_n \lambda}{\mu} \cdot \pi(n)$$

If N is large enough and $n \geq N$, then $a_n \lambda / \mu \leq 1/2$ and it follows that

$$\pi(n+1) \leq \left(\frac{1}{2}\right)^{n-N} \pi(N)$$

This implies that $\sum_n \pi(n) < \infty$, so we can pick $\pi(0)$ to make the sum $= 1$. $\square$

Concrete Example. Suppose $a_n = 1/(n+1)$. In this case

$$\frac{\lambda_{n-1} \cdots \lambda_0}{\mu_n \cdots \mu_1} = \frac{\lambda^n}{\mu^n} \cdot \frac{1}{1 \cdot 2 \cdots n} = \frac{(\lambda/\mu)^n}{n!}$$

To find the stationary distribution we want to take $\pi(0) = c$ so that

$$c \sum_{n=0}^{\infty} \frac{(\lambda/\mu)^n}{n!} = 1$$

Recalling the formula for the Poisson distribution with mean λ/μ, we see that $c = e^{-\lambda/\mu}$ and the stationary distribution is Poisson.

4.5.2 Multiple Servers

Example 4.27 (M/M/s Queue). Imagine a bank with $s \geq 1$ tellers that serve customers who queue in a single line if all servers are busy. We imagine that customers arrive at the times of a Poisson process with rate λ, and each requires an independent amount of service that has an exponential distribution with rate μ. As explained in Example 1.3, the flip rates are $q(n, n+1) = \lambda$ and

$$q(n, n-1) = \begin{cases} \mu n & \text{if } n \leq s \\ \mu s & \text{if } n \geq s \end{cases}$$

The conditions that result from using the detailed balance condition are

$$\lambda \pi(j-1) = \mu j \pi(j) \quad \text{for } j \leq s,$$
$$\lambda \pi(j-1) = \mu j \pi(j) \quad \text{for } j \geq s.$$

From this we conclude that

$$\pi(k) = \begin{cases} \dfrac{c}{k!} \left(\dfrac{\lambda}{\mu}\right)^k & k \leq s \\[2ex] \dfrac{c}{s! s^{k-s}} \left(\dfrac{\lambda}{\mu}\right)^k & k \geq s \end{cases} \tag{4.38}$$

where c is a constant that makes the sum equal to 1. From the last formula we see that if $\lambda < s\mu$, then $\sum_{j=0}^{\infty} \pi(j) < \infty$ and it is possible to pick c to make the sum equal to 1.

The condition $\lambda < s\mu$ for the existence of a stationary distribution is natural since it says that the service rate of the fully loaded system is larger than the arrival rate, so the queue will not grow out of control. This establishes the first result in

Theorem 4.12. *If* $\lambda < s\mu$, *then the* $M/M/s$ *queue is positive recurrent.*
If $\lambda = s\mu$, *then the* $M/M/s$ *queue is null recurrent.*
If $\lambda > s\mu$, *then the* $M/M/s$ *queue is transient.*

Proof. To prove the last two conclusions we note that an $M/M/s$ queue with s rate μ servers is less efficient than an $M/M/1$ queue with 1 rate $s\mu$ server, since the single server queue always has departures at rate $s\mu$, while the s server queue sometimes has departures at rate $n\mu$ with $n < s$. An $M/M/1$ queue is transient if its arrival rate is larger than its service rate, so the $M/M/s$ queue must be transient if $\lambda > s\mu$. The last reasoning implies that when $\lambda = \mu s$ the $M/M/s$ cannot be positive recurrent. To see that it is recurrent, note that the transition rates of the $M/M/s$ and $M/M/1$ queues agree for $n \geq s$, so $P_m(T_s < \infty) = 1$ for $m > s$ and hence $P_m(T_0 < \infty) = 1$. □

Formulas for the stationary distribution $\pi(n)$ for the $M/M/s$ queue are unpleasant to write down for a general number of servers s, but it is not hard to use (4.38) to find the stationary distribution in a concrete cases: If $s = 3$, $\lambda = 2$, and $\mu = 1$, then

$$\sum_{k=2}^{\infty} \pi(k) = \frac{c}{2} \cdot 2^2 \sum_{j=0}^{\infty} (2/3)^j = 6c$$

so $\sum_{k=0}^{\infty} \pi(k) = 9c$ and we have

$$\pi(0) = \frac{1}{9}, \quad \pi(1) = \frac{2}{9}, \quad \pi(k) = \frac{2}{9}\left(\frac{2}{3}\right)^{k-2} \quad \text{for } k \geq 2$$

4.5.3   Departure Processes

Our next result is a remarkable property of the M/M/s queue.

Theorem 4.13. *If* $\lambda < \mu s$, *then the output process of the M/M/s queue in equilibrium is a rate* λ *Poisson process.*

Your first reaction to this should be that it is crazy. Customers depart at rate 0, μ, 2μ, ..., $s\mu$, depending on the number of servers that are busy and it is usually the case that none of these numbers $= \lambda$. To further emphasize the surprising nature of Theorem 4.13, suppose for concreteness that there is one server, $\lambda = 1$, and $\mu = 10$. If, in this situation, we have just seen 30 departures in the last two hours, then it seems reasonable to guess that the server is busy and the next departure will be exponential(10). However, if the output process is Poisson, then the number of departures in disjoint intervals are independent.

Proof of Theorem 4.13. By repeating the proof of (1.13) one can show

Lemma 4.14. *Fix T and let* $Y_s = X_{T-s}$ *for* $0 \leq s \leq T$. *Then* Y_s *is a Markov chain with transition probability*

$$\hat{p}_t(i,j) = \frac{\pi(j)p_t(j,i)}{\pi(i)}$$

Proof. If $s + t \leq T$, then

$$P(Y_{s+t} = j|Y_s = i) = \frac{P(Y_{s+t} = j, Y_s = i)}{P(Y_s = i)} = \frac{P(X_{T-(s+t)} = j, X_{T-s} = i)}{P(X_{T-s} = i)}$$

$$= \frac{P(X_{T-(s+t)} = j)P(X_{T-s} = i|X_{T-(s+t)} = j)}{\pi(i)} = \frac{\pi(j)p_t(j,i)}{\pi(i)}$$

which is the desired result.   □

If π satisfies the detailed balance condition $\pi(i)q(i,j) = \pi(j)q(j,i)$, then the reversed chain has transition probability $\hat{p}_t(i,j) = p_t(i,j)$.

As we learned in Example 4.27, when $\lambda < \mu s$ the $M/M/s$ queue is a birth and death chain with a stationary distribution π that satisfies the detailed balance condition. Lemma 4.14 implies that if we take the movie of the Markov chain in equilibrium and run it backwards, then we see something that has the same distribution as the $M/M/s$ queue. Reversing time turns arrivals into departures, so the departures must be a Poisson process with rate λ.   □

It should be clear from the proof just given that we also have

Theorem 4.15. *Consider a queue in which arrivals occur according to a Poisson process with rate λ and customers are served at rate μ_n when there are n in the system. Then as long as there is a stationary distribution the output process will be a rate λ Poisson process.*

A second refinement that will be useful in the next section is

Theorem 4.16. *Let $N(t)$ be the number of departures between time 0 and time t for the M/M/1 queue $X(t)$ started from its equilibrium distribution. Then $\{N(s) : 0 \leq s \leq t\}$ and $X(t)$ are independent.*

Why is this true? At first it may sound deranged to claim that the output process up to time t is independent of the queue length. However, if we reverse time, then the departures before time t turn into arrivals after t, and these are obviously independent of the queue length at time t, $X(t)$.   □

4.6   Queueing Networks*

In many situations we are confronted with more than one queue. For example, when you go to the Department of Motor Vehicles in North Carolina to renew your driver's license you must (i) get your vision and sign recognition tested, (ii) pay your fees, and (iii) get your picture taken. A simple model of this type of situation with only two steps is

Example 4.28 (Two-Station Tandem Queue). In this system customers at times of a Poisson process with rate λ arrive at service facility 1 where they each require an independent exponential amount of service with rate μ_1. When they complete service at the first site, they join a second queue to wait for an exponential amount of service with rate μ_2.

Our main problem is to find conditions that guarantee that the queue stabilizes, i.e., has a stationary distribution. This is simple in the tandem queue. The first queue is not affected by the second, so if $\lambda < \mu_1$, then (4.35) tells us that the equilibrium probability of the number of customers in the first queue, X_t^1, is given by the shifted geometric distribution

$$P(X_t^1 = m) = \left(\frac{\lambda}{\mu_1}\right)^m \left(1 - \frac{\lambda}{\mu_1}\right)$$

In the previous section we have shown that the output process of an $M/M/1$ queue in equilibrium is a rate λ Poisson process. This means that if the first queue is in equilibrium, then the number of customers in the queue, X_t^2, is itself an $M/M/1$ queue with arrivals at rate λ (the output rate for 1) and service rate μ_2. Using the results in (4.35) again, the number of individuals in the second queue has stationary distribution

$$P(X_t^2 = n) = \left(\frac{\lambda}{\mu_2}\right)^n \left(1 - \frac{\lambda}{\mu_2}\right)$$

To specify the stationary distribution of the system, we need to know the joint distribution of X_t^1 and X_t^2. The answer is somewhat remarkable: in equilibrium the two queue lengths are independent.

$$P(X_t^1 = m, X_t^2 = n) = \left(\frac{\lambda}{\mu_1}\right)^m \left(1 - \frac{\lambda}{\mu_1}\right) \cdot \left(\frac{\lambda}{\mu_2}\right)^n \left(1 - \frac{\lambda}{\mu_2}\right) \tag{4.39}$$

Why is this true? Theorem 4.16 implies that the queue length and the departure process are independent.

Since there is more than a little hand-waving going on in the proof of Theorem 4.16 and its application here, it is comforting to note that one can simply verify from the definitions that

Lemma 4.17. *If* $\pi(m, n) = c\lambda^{m+n}/(\mu_1^m \mu_2^n)$, *where* $c = (1 - \lambda/\mu_1)(1 - \lambda/\mu_2)$ *is a constant chosen to make the probabilities sum to 1, then* π *is a stationary distribution.*

Proof. The first step in checking $\pi Q = 0$ is to compute the rate matrix Q. To do this it is useful to draw a picture which assumes $m, n > 0$

The rate arrows plus the ordinary lines on the picture make three triangles. We will now check that the flows out of and into (m, n) in each triangle balance. In symbols we note that

$$(a) \qquad \mu_1 \pi(m, n) = \frac{c\lambda^{m+n}}{\mu_1^{m-1}\mu_2^n} = \lambda\pi(m-1, n)$$

$$(b) \qquad \mu_2 \pi(m, n) = \frac{c\lambda^{m+n}}{\mu_1^m \mu_2^{n-1}} = \mu_1\pi(m+1, n-1)$$

$$(c) \qquad \lambda\pi(m, n) = \frac{c\lambda^{m+n+1}}{\mu_1^m \mu_2^n} = \mu_2\pi(m, n+1)$$

This shows that $\pi Q = 0$ when $m, n > 0$. There are three other cases to consider: (i) $m = 0, n > 0$, (ii) $m > 0, n = 0$, and (iii) $m = 0, n = 0$. In these cases some of the rates are missing: (i) those in (a), (ii) those in (b), and (iii) those in (a) and (b). However, since the rates in each group balance we have $\pi Q = 0$.   □

Example 4.29 (General Two-Station Queue). Suppose that at station i: arrivals from outside the system occur at rate λ_i, service occurs at rate μ_i, and departures go to the other queue with probability p_i and leave the system with probability $1 - p_i$.

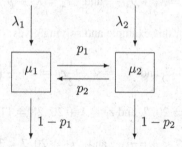

Our question is: When is the system stable? That is, when is there a stationary distribution? To get started on this question suppose that both servers are busy. In this case work arrives at station 1 at rate $\lambda_1 + p_2\mu_2$, and work arrives at station 2 at rate $\lambda_2 + p_1\mu_1$. It should be intuitively clear that:

(i) if $\lambda_1 + p_2\mu_2 < \mu_1$ and $\lambda_2 + p_1\mu_1 < \mu_2$, then each server can handle their maximum arrival rate and the system will have a stationary distribution.
(ii) if $\lambda_1 + p_2\mu_2 > \mu_1$ and $\lambda_2 + p_1\mu_1 > \mu_2$, then there is positive probability that both servers will stay busy for all time and the queue lengths will tend to infinity.

Not covered by (i) or (ii) is the situation in which server 1 can handle her worst case scenario but server 2 cannot cope with his:

$$\lambda_1 + p_2\mu_2 < \mu_1 \quad \text{and} \quad \lambda_2 + p_1\mu_1 > \mu_2$$

In some situations in this case, queue 1 will be empty often enough to reduce the arrivals at station 2 so that server 2 can cope with his workload. As we will see, a concrete example of this phenomenon occurs when

$$\lambda_1 = 1, \quad \mu_1 = 4, \quad p_1 = 1/2 \quad \lambda_2 = 2, \quad \mu_2 = 3.5, \quad p_2 = 1/4$$

To check that for these rates server 1 can handle the maximum arrival rate but server 2 cannot, we note that

$$\lambda_1 + p_2\mu_2 = 1 + \frac{1}{4} \cdot 3.5 = 1.875 < 4 = \mu_1$$

$$\lambda_2 + p_1\mu_1 = 2 + \frac{1}{2} \cdot 4 = 4 > 3.5 = \mu_2$$

To derive general conditions that will allow us to determine when a two-station network is stable, let r_i be the long run average rate that customers arrive at station i. If there is a stationary distribution, then r_i must also be the long run average rate at which customers leave station i or the queue would grow linearly in time. If we want the flow in and out of each of the stations to balance, then we need

$$r_1 = \lambda_1 + p_2 r_2 \quad \text{and} \quad r_2 = \lambda_2 + p_1 r_1 \tag{4.40}$$

Plugging in the values for this example and solving gives

$$r_1 = 1 + \frac{1}{4}r_2 \quad \text{and} \quad r_2 = 2 + \frac{1}{2}r_1 = 2 + \frac{1}{2}\left(1 + \frac{1}{4}r_2\right)$$

So $(7/8)r_2 = 5/2$ or $r_2 = 20/7$, and $r_1 = 1 + 20/28 = 11/7$. Since

$$r_1 = 11/7 < 4 = \mu_1 \quad \text{and} \quad r_2 = 20/7 < 3.5 = \mu_2$$

this analysis suggests that there will be a stationary distribution.

To prove that there is one, we return to the general situation and suppose that the r_i we find from solving (4.40) satisfy $r_i < \mu_i$. Thinking of two independent $M/M/1$ queues with arrival rates r_i, we let $\alpha_i = r_i/\mu_i$ and guess:

Theorem 4.18. *If* $\pi(m, n) = c\alpha_1^m\alpha_2^n$ *where* $c = (1 - \alpha_1)(1 - \alpha_2)$, *then* π *is a stationary distribution.*

Proof. The first step in checking $\pi Q = 0$ is to compute the rate matrix Q. To do this it is useful to draw a picture. Here, we have assumed that m and n are both positive. To make the picture slightly less cluttered, we have only labeled half of the arrows and have used $q_i = 1 - p_i$.

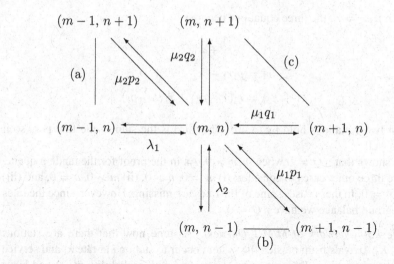

The rate arrows plus the dotted lines in the picture make three triangles. We will now check that the flows out of and into (m, n) in each triangle balance. In symbols we need to show that

(a) $\quad \mu_1 \pi(m, n) = \mu_2 p_2 \pi(m - 1, n + 1) + \lambda_1 \pi(m - 1, n)$

(b) $\quad \mu_2 \pi(m, n) = \mu_1 p_1 \pi(m + 1, n - 1) + \lambda_2 \pi(m, n - 1)$

(c) $\quad (\lambda_1 + \lambda_2) \pi(m, n) = \mu_2 (1 - p_2) \pi(m, n + 1) + \mu_1 (1 - p_1) \pi(m + 1, n)$

Filling in $\pi(m, n) = c \alpha_1^m \alpha_2^n$ and canceling out c, we have

$$\mu_1 \alpha_1^m \alpha_2^n = \mu_2 p_2 \alpha_1^{m-1} \alpha_2^{n+1} + \lambda_1 \alpha_1^{m-1} \alpha_2^n$$

$$\mu_2 \alpha_1^m \alpha_2^n = \mu_1 p_1 \alpha_1^{m+1} \alpha_2^{n-1} + \lambda_2 \alpha_1^m \alpha_2^{n-1}$$

$$(\lambda_1 + \lambda_2) \alpha_1^m \alpha_2^n = \mu_2 (1 - p_2) \alpha_1^m \alpha_2^{n+1} + \mu_1 (1 - p_1) \alpha_1^{m+1} \alpha_2^n$$

Canceling out the highest powers of α_1 and α_2 common to all terms in each equation gives

$$\mu_1 \alpha_1 = \mu_2 p_2 \alpha_2 + \lambda_1$$

$$\mu_2 \alpha_2 = \mu_1 p_1 \alpha_1 + \lambda_2$$

$$(\lambda_1 + \lambda_2) = \mu_2 (1 - p_2) \alpha_2 + \mu_1 (1 - p_1) \alpha_1$$

Filling in $\mu_i \alpha_i = r_i$, the three equations become

$$r_1 = p_2 r_2 + \lambda_1$$
$$r_2 = p_1 r_1 + \lambda_2$$
$$(\lambda_1 + \lambda_2) = r_2(1 - p_2) + r_1(1 - p_1)$$

The first two equations hold by (4.40). The third is the sum of the first two, so it holds as well.

This shows that $\pi Q = 0$ when $m, n > 0$. As in the proof for the tandem queue, there are three other cases to consider: (i) $m = 0, n > 0$, (ii) $m > 0, n = 0$, and (iii) $m = 0, n = 0$. In these cases some of the rates are missing. However, since the rates in each group balance we have $\pi Q = 0$.   □

Example 4.30 (Network of M/M/1 Queues). Assume now that there are stations $1 \leq i \leq K$. Arrivals from outside the system occur to station i at rate λ_i and service occurs there at rate μ_i. Departures go to station j with probability $p(i,j)$ and leave the system with probability

$$q(i) = 1 - \sum_j p(i,j) \qquad (4.41)$$

To have a chance of stability we must suppose

(A)   *For each i it is possible for a customer entering at i to leave the system. That is, for each i there is a sequence of states $i = j_0, j_1, \ldots j_n$ with $p(j_{m-1}, j_m) > 0$ for $1 \leq m \leq n$ and $q(j_n) > 0$.*

Generalizing (4.40), we investigate stability by solving the system of equations for the r_j that represent the arrival rate at station j. As remarked earlier, the departure rate from station j must equal the arrival rate, or a linearly growing queue would develop. Thinking about the arrival rate at j in two different ways, it follows that

$$r_j = \lambda_j + \sum_{i=1}^{K} r_i p(i,j) \qquad (4.42)$$

This equation can be rewritten in matrix form as $r = \lambda + rp$ and solved as

$$r = \lambda(I - p)^{-1} \qquad (4.43)$$

By reasoning in Sect. 1.9, where unfortunately r is what we are calling p here:

$$r = \sum_{n=0}^{\infty} \lambda p^n = \sum_{n=0}^{\infty} \sum_{i=1}^{K} \lambda_i p^n(i,j)$$

The answer is reasonable: $p^n(i,j)$ is the probability a customer entering at i is at j after he has completed n services. The sum then adds the rates for all the ways of arriving at j.

Having found the arrival rates at each station, we can again be brave and guess that if $r_j < \mu_j$, then the stationary distribution is given by

$$\pi(n_1,\ldots,n_K) = \prod_{j=1}^{K} \left(\frac{r_j}{\mu_j}\right)^{n_j} \left(1 - \frac{r_j}{\mu_j}\right) \tag{4.44}$$

This is true, but the proof is more complicated than for the two-station examples, so we omit it.

Example 4.31. At a government agency entering customers always go to server 1. After completing service there, 30 % leave the system while 70 % go to server 2. At server 2, 50 % go to server 3, 20 % of the customers have to return to server 1, and 30 % leave the system. From server 3, 20 % go back to server 2 but the other 80 % can go. That is, the routing matrix is

$$p = \begin{array}{c} \\ 1 \\ 2 \\ 3 \end{array} \begin{array}{ccc} 1 & 2 & 3 \\ 0 & .7 & 0 \\ .2 & 0 & .5 \\ 0 & .2 & 0 \end{array}$$

Suppose that arrivals from outside only occur at server 1 and at rate $\lambda_1 = 3.8$ per hour. Find the stationary distribution if the service rates are $\mu_1 = 9$, $\mu_2 = 7$, and $\mu_3 = 7$.

The first step is to solve the equations

$$r_j = \lambda_j + \sum_{i=1}^{3} r_i p(i,j)$$

By (4.42) the solution is $r = \lambda(I - p)^{-1}$, where

$$(I - p)^{-1} = \begin{pmatrix} 45/38 & 35/38 & 35/76 \\ 5/19 & 25/19 & 25/38 \\ 1/19 & 5/19 & 43/38 \end{pmatrix}$$

so multiplying the first row by λ_1, we have $r_1 = 9/2$, $r_2 = 7/2$, and $r_3 = 7/4$. Since each $r_i < \mu_i$ the stationary distribution is

$$\pi(n_1, n_2, n_3) = \frac{3}{16}(1/2)^{n_1}(1/2)^{n_2}(1/4)^{n_3}$$

It is easy to see from the proof that Little's formula also applies to queueing networks. In this case the average number of people in the system is

$$L = \sum_{i=1}^{3} \frac{1}{1 - (r_i/\mu_i)} - 1 = 1 + 1 + \frac{1}{3} = \frac{7}{3}$$

so the average waiting time for a customer entering the system is

$$W = \frac{L}{\lambda} = \frac{7/3}{19/5} = \frac{35}{57} = 0.6140.$$

4.7 Chapter Summary

In principle continuous time Markov chains are defined by giving their transition probabilities $p_t(i,j)$, which satisfy the Chapman–Kolmogorov equation.

$$\sum_k p_s(i,k)p_t(k,j) = p_{s+t}(i,j)$$

In practice, the basic data to describe the chain are the rates $q(i,j)$ at which jumps occur from i to $j \neq i$. If we let $\lambda_i = \sum_{j \neq i} q(i,j)$ be the total rate of jumps out of i and let

$$Q(i,j) = \begin{cases} q(i,j) & i \neq j \\ -\lambda_i & i = j \end{cases}$$

then the transition probability satisfies the Kolmogorov differential equations:

$$p_t'(i,j) = \sum_k Q(i,k)p_t(k,j) = \sum_k p_t(i,k)Q(k,j)$$

These equations can only be solved explicitly in a small number of examples, but they are essential for developing the theory.

Embedded Markov Chain The discrete time chain with transition probability

$$r(i,j) = \frac{q(i,j)}{\lambda_i}$$

goes through the same sequence of states as X_t but stays in each one for one unit of time.

Stationary Distributions A stationary distribution has $\sum_i \pi(i) = 1$ and satisfies $\pi p_t = \pi$ for all $t > 0$, which is equivalent to $\pi Q = 0$. To solve these equations mechanically, we replace the last column of Q by all 1's to define a matrix A and then π will be the last row of A^{-1}.

If X_t is irreducible and has stationary distribution π, then

$$p_t(i,j) \to \pi(j) \quad \text{as } t \to \infty$$

Detailed Balance Condition A sufficient condition to be stationary is that

$$\pi(i)q(i,j) = \pi(j)q(j,i)$$

There may not be a stationary distribution with this property, but there is one if we have a birth and death chain: i.e., the state space is $\{0, 1, \ldots r\}$, where r may be ∞, and we have $q(i,j) = 0$ when $|i - j| > 1$. In this case we have

$$\pi(n) = \frac{\lambda_{n-1} \cdots \lambda_0}{\mu_n \cdots \mu_1} \cdot \pi(0)$$

Queues provide a number of interesting examples of birth and death chains.

Exit Distribution Let $V_A = \min\{t : X_t \in A\}$ be the time of the first visit to A. Let A and B be sets so that $C = S - (A \cup B)$ is finite and for each $x \in C$, $P_x(V_A \wedge V_B < \infty) > 0$ Then $h(x) = P_x(V_A < V_B)$ satisfies $h(a) = 1$ for $a \in A$, $h(b) = 0$ for $b \in B$ and

$$h(x) = \sum_y r(x,y)h(y) \quad \text{for } x \in C.$$

One can also work directly with the transition rates. The boundary conditions are the same, while

$$\sum_x Q(x,y)h(y) = 0 \quad \text{for } x \in C$$

If we define a column vector by $v(i) = \sum_{j \in A} q(i,j)$ for $i \in C$, then the solution is

$$h = (-R)^{-1}v$$

Hitting Times The expected hitting time $g(x) = E_x V_A$ satisfies $g(a) = 0$ for $a \in A$ and

$$g(x) = \frac{1}{\lambda_x} + \sum_j r(x,y)g(y) \quad \text{for } x \notin A.$$

If we let R be the part of the Q-matrix where $i, j \notin A$, then

$$g = -R^{-1}\mathbf{1}$$

where $\mathbf{1}$ is a column vector of all 1's.

Busy Periods In queueing chains we can compute the duration of the busy period $E_1 T_0$ from

$$\pi(0) = \frac{1/\lambda_0}{1/\lambda_0 + E_1 T_0}$$

Solving we have $1 + \lambda_0 E_1 T_0 = 1/\pi(0)$ and hence

$$E_1 T_0 = \frac{1 - \pi(0)}{\lambda \pi(0)} \tag{4.45}$$

4.8 Exercises

4.1. A salesman flies around between Atlanta, Boston, and Chicago as the following rates (the units are trips per month):

$$
\begin{array}{c c c c}
 & A & B & C \\
A & -4 & 2 & 2 \\
B & 3 & -4 & 1 \\
C & 5 & 0 & -5 \\
\end{array}
$$

(a) Find the limiting fraction of time she spends in each city. (b) What is her average number of trips each year from Boston to Atlanta?

4.2. A hemoglobin molecule can carry one oxygen or one carbon monoxide molecule. Suppose that the two types of gases arrive at rates 1 and 2 and attach for an exponential amount of time with rates 3 and 4, respectively. Formulate a Markov chain model with state space $\{+, 0, -\}$ where $+$ denotes an attached oxygen molecule, $-$ an attached carbon monoxide molecule, and 0 a free hemoglobin molecule and find the long run fraction of time the hemoglobin molecule is in each of its three states.

4.3. A machine is subject to failures of types $i = 1, 2, 3$ at rates $\lambda_1 = 1/24$, $\lambda_2 = 1/30$, $\lambda_3 = 1/84$. A failure of type i takes an exponential amount of time with rate $\mu_1 = 1/3$, $\mu_2 = 1/5$, and $\mu_3 = 1/7$. Formulate a Markov chain model with state space $\{0, 1, 2, 3\}$ and find its stationary distribution.

4.4. A small computer store has room to display up to three computers for sale. Customers come at times of a Poisson process with rate 2 per week to buy a computer and will buy one if at least 1 is available. When the store has only one computer left it places an order for two more computers. The order takes an exponentially distributed amount of time with mean 1 week to arrive. Of course, while the store is waiting for delivery, sales may reduce the inventory to 1 and then to 0. (a) Write down the matrix of transition rates Q_{ij} and solve $\pi Q = 0$ to find the stationary distribution. (b) At what rate does the store make sales?

4.5. Consider two machines that are maintained by a single repairman. Machine i functions for an exponentially distributed amount of time with rate λ_i before it fails. The repair times for each unit are exponential with rate μ_i. They are repaired in the order in which they fail. (a) Formulate a Markov chain model for this situation with state space $\{0, 1, 2, 12, 21\}$. (b) Suppose that $\lambda_1 = 1$, $\mu_1 = 2$, $\lambda_2 = 3$, $\mu_2 = 4$. Find the stationary distribution.

4.6. Consider the set-up of the previous problem but now suppose machine 1 is much more important than 2, so the repairman will always service 1 if it is broken. (a) Formulate a Markov chain model for the this system with state space $\{0, 1, 2, 12\}$ where the numbers indicate the machines that are broken at the time. (b) Suppose that $\lambda_1 = 1$, $\mu_1 = 2$, $\lambda_2 = 3$, $\mu_2 = 4$. Find the stationary distribution.

4.7. Two people are working in a small office selling shares in a mutual fund. Each is either on the phone or not. Suppose that calls come in to the two brokers at rate $\lambda_1 = \lambda_2 = 1$ per hour, while the calls are serviced at rate $\mu_1 = \mu_2 = 3$. (a) Formulate a Markov chain model for this system with state space $\{0, 1, 2, 12\}$ where the state indicates who is on the phone. (b) Find the stationary distribution. (c) Suppose they upgrade their telephone system so that a call to one line that is busy is forwarded to the other phone and lost if that phone is busy. Find the new stationary probabilities. (d) Compare the rate at which calls are lost in the two systems.

4.8. Customers arrive at a full-service one-pump gas station at rate of 20 cars per hour. However, customers will go to another station if there are at least two cars in the station, i.e., one being served and one waiting. Suppose that the service time for customers is exponential with mean six minutes. (a) Formulate a Markov chain model for the number of cars at the gas station and find its stationary distribution. (b) On the average how many customers are served per hour?

4.9. Solve the previous problem for a two-pump self-serve station under the assumption that customers will go to another station if there are at least four cars in the station, i.e., two being served and two waiting.

4.10. A computer lab has three laser printers, two that are hooked to the network and one that is used as a spare. A working printer will function for an exponential amount of time with mean 20 days. Upon failure it is immediately sent to the repair facility and replaced by another machine if there is one in working order. At the repair facility machines are worked on by a single repairman who needs an exponentially distributed amount of time with mean 2 days to fix one printer. In the long run how often are there two working printers?

4.11. A computer lab has three laser printers that are hooked to the network. A working printer will function for an exponential amount of time with mean 20 days. Upon failure it is immediately sent to the repair facility. There machines are worked on by two repairman who can each repair one printer in an exponential amount of time with mean 2 days. However, it is not possible for two people to work on one printer at once. (a) Formulate a Markov chain model for the number of working printers and find the stationary distribution. (b) How often are both repairmen busy? (c) What is the average number of machines in use?

4.12. A computer lab has three laser printers and five toner cartridges. Each machine requires one toner cartridges which lasts for an exponentially distributed amount of time with mean 6 days. When a toner cartridge is empty it is sent to a repairman who takes an exponential amount of time with mean 1 day to refill it. (a) Compute the stationary distribution. (b) How often are all three printers working?

4.13. A taxi company has three cabs. Calls come in to the dispatcher at times of a Poisson process with rate 2 per hour. Suppose that each requires an exponential amount of time with mean 20 minutes, and that callers will hang up if they hear there are no cabs available. (a) What is the probability all three cabs are busy when a call comes in? (b) In the long run, on the average how many customers are served per hour?

4.14. A small company maintains a fleet of four cars to be driven by its workers on business trips. Requests to use cars are a Poisson process with rate 1.5 per day. A car is used for an exponentially distributed time with mean 2 days. Forgetting about weekends, we arrive at the following Markov chain for the number of cars in service:

	0	1	2	3	4
0	-1.5	1.5	0	0	0
1	0.5	-2.0	1.5	0	0
2	0	1.0	-2.5	1.5	0
3	0	0	1.5	-3	1.5
4	0	0	0	2	-2

(a) Find the stationary distribution. (b) At what rate do unfulfilled requests come in? How would this change if there were only three cars? (c) Let $g(i) = E_i T_4$. Write and solve equations to find the $g(i)$. (d) Use the stationary distribution to compute $E_3 T_4$.

Hitting Times and Exit Distributions

4.15. Consider the salesman from Problem 4.1. She just left Atlanta. (a) What is the expected time until she returns to Atlanta? (b) Find the answer to (a) by computing the stationary distribution.

4.16. Brad's relationship with his girl friend Angelina changes between Amorous, Bickering, Confusion, and Depression according to the following transition rates when t is the time in months:

$$
\begin{array}{c c c c c}
  & \text{A} & \text{B} & \text{C} & \text{D} \\
\text{A} & -4 & 3 & 1 & 0 \\
\text{B} & 4 & -6 & 2 & 0 \\
\text{C} & 2 & 3 & -6 & 1 \\
\text{D} & 0 & 0 & 2 & -2
\end{array}
$$

(a) Find the long run fraction of time he spends in these four states? (b) Does the chain satisfy the detailed balance condition? (c) They are amorous now. What is the expected amount of time until depression sets in?

4.17. A submarine has three navigational devices but can remain at sea if at least two are working. Suppose that the failure times are exponential with means 1 year, 1.5 years, and 3 years. Formulate a Markov chain with states $0 =$ all parts working, 1, 2, 3 = one part failed, and 4 = two failures. Compute E_0T_4 to determine the average length of time the boat can remain at sea.

4.18. Al, Betty, Charlie, and Diane are working at the math club table during freshmen move-in week. Their attention spans for doing the job are independent and have exponential distributions with means 1/2, 2/3, 1, and 2 hours. (Rates are 2, 3/2, 1, 1/2.) (a) What is the expected amount of time until only two people are left? (b, 15 points) What is the probability the last two left have the same sex, i.e., AC or BD?

4.19. Excited by the recent warm weather Jill and Kelly are doing spring cleaning at their apartment. Jill takes an exponentially distributed amount of time with mean 30 minutes to clean the kitchen. Kelly takes an exponentially distributed amount of time with mean 40 minutes to clean the bath room. The first one to complete their task will go outside and start raking leaves, a task that takes an exponentially distributed amount of time with a mean of one hour. When the second person is done inside, they will help the other and raking will be done at rate 2. (Of course the other person may already be done raking in which case the chores are done.) What is the expected time until the chores are all done?

Markovian Queues: Finite State Space

4.20. Consider an $M/M/4$ queue with no waiting room, for example, four prostitutes hanging out on a street corner. Customers arrive at times of a Poisson process with rate 4. Each service takes an exponentially distributed amount of time with mean 1/2. If no server is free, then the customer goes away never to come back. (a) Find the stationary distribution. (b) At what rate do customers enter the system? (c) Use $W = L/\lambda_a$ to calculate the average time a customer spends in the system.

4.21. *Two queues in series.* Consider a two-station queueing network in which arrivals only occur at the first server and do so at rate 2. If a customer finds server 1 free he enters the system; otherwise he goes away. When a customer is done at the first server he moves on to the second server if it is free and leaves the system if it is not. Suppose that server 1 serves at rate 4 while server 2 serves at rate 2. Formulate a Markov chain model for this system with state space $\{0, 1, 2, 12\}$ where the state indicates the servers who are busy. In the long run (a) what proportion of customers enter the system? (b) What proportion of the customers visit server 2? (c) Compute $E_x T_0$ for $x = 1, 2, 12$ and use this to determine the expected duration of the busy period $E_1 T_0$. (d) Compute $E_1 T_0$ from $\pi(0)$ by using our result for alternating renewal processes.

4.22. Two people who prepare tax forms are working in a store at a local mall. Each has a chair next to his desk where customers can sit and be served. In addition there is one chair where customers can sit and wait. Customers arrive at rate 4 but will go away if there is already someone sitting in the chair waiting. Suppose that server i requires an exponential amount of time with rate μ_i, where $\mu_1 = 3$ and $\mu_2 = 5$ and that when both servers are free an arriving customer is equally likely to choose either one. (a) Formulate a Markov chain model for this system with state space $\{0, 1, 2, 12, 3\}$ where the first four states indicate the servers that are busy while the last indicates that there is a total of three customers in the system: one at each server and one waiting. (b) Find the stationary distribution. (c) At what rate do customers enter the system? (d) How much time does the average customer spend in the system? (e) What fraction of customers are served by server 1?

4.23. There are two tennis courts. Pairs of players arrive at rate 3 per hour and play for an exponentially distributed amount of time with mean one hour. If there are already two pairs of players waiting new arrivals will leave. (a) Find the stationary distribution for the number of courts occupied. (b) Find the rate at which customers enter the system. (c) Find the expected amount of time a pair has to wait before they can begin playing.

4.24. Carrboro High School is having a car wash to raise money for a Spring Break trip to Amsterdam. Eight students are working at the car wash. Jill and Kelly are on the sidewalk holding signs advertising the car wash. If there are one, two, or three cars getting washed, then the other six students divide themselves up between the cars. If there are four cars, then Jill and Kelly also help, so four cars are being washed by two people each. A car being washed by m students departs at rate m per hour. Cars arrive at rate 12 per hour and enter service with probability $1 - k/4$ if k cars are being washed. In short we have the following Markov chain for the number of cars being washed:

$$
\begin{array}{c|ccccc}
 & 0 & 1 & 2 & 3 & 4 \\
\hline
0 & -12 & 12 & & & \\
1 & 6 & -15 & 9 & & \\
2 & & 6 & -12 & 6 & \\
3 & & & 6 & -9 & 3 \\
4 & & & & 8 & -8 \\
\end{array}
$$

(a) Find the stationary distribution. (b) Compute the average number of cars in the system L, and the average time a car takes to get washed W. (c) Use $\pi(0)$ to compute the expected duration of the busy period $E_1 T_0$. (d) Find $E_1 T_0$ by computing $E_x T_0$ for $x = 1, 2, 3$.

4.25. Two teaching assistants work in a homework help room. They can each serve up to two students, but get less efficient when there are helping two at once. Students arrive at rate 3 but will not go in the room if there are already four students there. To cut a long story short the number of students in the help room is a Markov chain with Q matrix.

$$
\begin{array}{c|ccccc}
 & 0 & 1 & 2 & 3 & 4 \\
\hline
0 & -3 & 3 & 0 & 0 & 0 \\
1 & 2 & -5 & 3 & 0 & 0 \\
2 & 0 & 4 & -7 & 3 & 0 \\
3 & 0 & 0 & 5 & -8 & 3 \\
4 & 0 & 0 & 0 & 6 & -6 \\
\end{array}
$$

(a) Use the detailed balance condition to find the stationary distribution. (b) Use the formula $L = \lambda_a W$ to find the average time W that a student spends in the help room. (c) Use $\pi(0)$ to compute the expected duration of the busy period $E_1 T_0$. (d) Find $E_1 T_0$ by computing $E_x T_0$ for $x = 1, 2, 3, 4$.

Markovian Queues: Infinite State Space

4.26. Consider a taxi station at an airport where taxis and (groups of) customers arrive at times of Poisson processes with rates 2 and 3 per minute. Suppose that a taxi will wait no matter how many other taxis are present. However, if an arriving person does not find a taxi waiting he leaves to find alternative transportation. (a) Find the proportion of arriving customers that get taxis. (b) Find the average number of taxis waiting.

4.27. *Queue with impatient customers.* Customers arrive at a single server at rate λ and require an exponential amount of service with rate μ. Customers waiting in line are impatient and if they are not in service they will leave at rate δ independent of their position in the queue. (a) Show that for any $\delta > 0$ the system has a stationary distribution. (b) Find the stationary distribution in the very special case in which $\delta = \mu$.

4.28. Customers arrive at the Shortstop convenience store at a rate of 20 per hour. When two or fewer customers are present in the checkout line, a single clerk works and the service time is three minutes. However, when there are three or more customers are present, an assistant comes over to bag up the groceries and reduces the service time to two minutes. Assuming the service times are exponentially distributed, find the stationary distribution.

4.29. Customers arrive at a two-server station according to a Poisson process with rate 4. Upon arriving they join a single queue to wait for the next available server. Suppose that the service times of the two servers are exponential with rates 2 and 3 and that a customer who arrives to find the system empty will go to each of the servers with probability 1/2. Formulate a Markov chain model for this system with state space $\{0, a, b, 2, 3, \ldots\}$ where the states give the number of customers in the system, with a or b indicating there is one customer at a or b, respectively. Show that this system satisfies detailed balance and find the stationary distribution.

4.30. At present the Economics department and the Sociology department each has one typist who can type 25 letters a day. Economics requires an average of 20 letters per day, while Sociology requires only average of 15. Assuming Poisson arrival and exponentially distributed typing times find (a) the average queue length and average waiting time in each departments (b) the average overall waiting time if they merge their resources to form a typing pool.

4.31. A professor receives requests for letters of recommendation according to a Poisson process with rate 8 per month. When k are on his desk he feels stressed and writes them at rate $2k$ (per month). (a) Find the stationary distribution of the number of letters to be written. (b) What is the average time a person has to wait before their letter is written. (c) He just received a request for a letter. What is the probability his queue will return to 0 before it hits 4, i.e., what is $P_1(T_0 < T_4)$.

Queueing Networks

4.32. Consider a production system consisting of a machine center followed by an inspection station. Arrivals from outside the system occur only at the machine center and follow a Poisson process with rate λ. The machine center and inspection station are each single server operations with rates μ_1 and μ_2. Suppose that each item independently passes inspection with probability p. When an object fails inspection it is sent to the machine center for reworking. Find the conditions on the parameters that are necessary for the system to have a stationary distribution.

4.33. Consider a three station queueing network in which arrivals to servers $i = 1, 2, 3$ occur at rates $3, 2, 1$, while service at stations $i = 1, 2, 3$ occurs at rates $4, 5, 6$. Suppose that the probability of going to j when exiting i, $p(i, j)$ is given by $p(1, 2) = 1/3, p(1, 3) = 1/3, p(2, 3) = 2/3$, and $p(i, j) = 0$ otherwise. Find the stationary distribution.

4.34. *Feed-forward queues.* Consider a k station queueing network in which arrivals to server i occur at rate λ_i and service at station i occurs at rate μ_i. We say that the queueing network is feed-forward if the probability of going from i to $j < i$ has $p(i, j) = 0$. Consider a general three station feed-forward queue. What conditions on the rates must be satisfied for a stationary distribution to exist?

4.35. *Queues in series.* Consider a k station queueing network in which arrivals to server i occur at rate λ_i and service at station i occurs at rate μ_i. In this problem we examine the special case of the feed-forward system in which $p(i, i + 1) = p_i$ for $1 \leq i < k$. In words the customer goes to the next station or leaves the system. What conditions on the rates must be satisfied for a stationary distribution to exist?

4.36. At registration at a very small college, students arrive at the English table at rate 10 and at the Math table at rate 5. A student who completes service at the English table goes to the Math table with probability 1/4 and to the cashier with probability 3/4. A student who completes service at the Math table goes to the English table with probability 2/5 and to the cashier with probability 3/5. Students who reach the cashier leave the system after they pay. Suppose that the service times for the English table, Math table, and cashier are 25, 30, and 20, respectively. Find the stationary distribution.

4.37. At a local grocery store there are queues for service at the fish counter (1), meat counter (2), and café (3). For $i = 1, 2, 3$ customers arrive from outside the system to station i at rate i, and receive service at rate $4 + i$. A customer leaving station i goes to j with probabilities $p(i, j)$ given the following matrix:

	1	2	3
1	0	1/4	1/2
2	1/5	0	1/5
3	1/3	1/3	0

In equilibrium what is the probability no one is in the system, i.e., $\pi(0, 0, 0)$.

4.38. Three vendors have vegetable stands in a row. Customers arrive at the stands 1, 2, and 3 at rates 10, 8, and 6. A customer visiting stand 1 buys something and leaves with probability 1/2 or visits stand 2 with probability 1/2. A customer visiting stand 3 buys something and leaves with probability 7/10 or visits stand 2 with probability 3/10. A customer visiting stand 2 buys something and leaves with probability 4/10 or visits stands 1 or 3 with probability 3/10 each. Suppose that the service rates at the three stands are large enough so that a stationary distribution exists. At what rate do the three stands make sales. To check your answer note that since each entering customers buys exactly once the three rates must add up to $10 + 8 + 6 = 24$.

4.39. Four children are playing two video games. The first game, which takes an average of four minutes to play, is not very exciting, so when a child completes a turn on it they always stand in line to play the other one. The second one, which takes an average of eight minutes, is more interesting so when they are done they will get back in line to play it with probability 1/2 or go to the other machine with probability 1/2. Assuming that the turns take an exponentially distributed amount of time, find the stationary distribution of the number of children playing or in line at each of the two machines.

Chapter 5
Martingales

In this chapter we will introduce a class of process that can be thought of as the fortune of a gambler betting on a fair game. These results will be important when we consider applications to finance in the next chapter. In addition, they will allow us to give more transparent proofs of some facts from Chap. 1 concerning exit distributions and exit times for Markov chains.

5.1 Conditional Expectation

Our study of martingales will rely heavily on the notion of conditional expectation and involve some formulas that may not be familiar, so we will review them here. We begin with several definitions. Given an event A we define its **indicator function**

$$1_A = \begin{cases} 1 & x \in A \\ 0 & x \in A^c \end{cases}$$

In words, 1_A is "1 on A" (and 0 otherwise). Given a random variable Y, we define the **integral of Y over A** to be

$$E(Y;A) = E(Y1_A)$$

Note that multiplying Y by 1_A sets the product $= 0$ on A^c and leaves the values on A unchanged. Finally, we define the **conditional expectation of Y given A** to be

$$E(Y|A) = E(Y;A)/P(A)$$

© Springer International Publishing Switzerland 2016
R. Durrett, *Essentials of Stochastic Processes*, Springer Texts in Statistics,
DOI 10.1007/978-3-319-45614-0_5

This is the expected value for the conditional probability defined by

$$P(\cdot|A) = P(\cdot \cap A)/P(A)$$

Example 5.1. A simple but important special case arises when the random variable Y and the set A are independent, i.e., for any set B we have

$$P(Y \in B, A) = P(Y \in B)P(A)$$

Noticing that this implies that $P(Y \in B, A^c) = P(Y \in B)P(A^c)$ and comparing with the definition of independence of random variables in (A.13), we see that this holds if and only if Y and 1_A are independent, so Theorem A.1 implies

$$E(Y; A) = E(Y1_A) = EY \cdot E1_A$$

and we have

$$E(Y|A) = EY \qquad\qquad\qquad (5.1)$$

It is easy to see from the definition that the integral over A is linear:

$$E(Y + Z; A) = E(Y; A) + E(Z; A) \qquad\qquad (5.2)$$

so dividing by $P(A)$, conditional expectation also has this property

$$E(Y + Z|A) = E(Y|A) + E(Z|A) \qquad\qquad (5.3)$$

Here and in later formulas and theorems, we always assume that all of the indicated expected values exist.

In addition, as in ordinary integration one can take "constants" outside of the integral. This property is very important for computations.

Lemma 5.1. *If X is a constant c on A, then $E(XY|A) = cE(Y|A)$.*

Proof. Since $X = c$ on A, $XY1_A = cY1_A$. Taking expected values and pulling the constant out front, $E(XY1_A) = E(cY1_A) = cE(Y1_A)$. Dividing by $P(A)$ now gives the result.   □

Being an expected value $E(\cdot|A)$ it has all of the usual properties, in particular:

Lemma 5.2 (Jensen's Inequality). *If ϕ is convex, then*

$$E(\phi(X)|A) \geq \phi(E(X|A))$$

Our next two properties concern the behavior of $E(Y; A)$ and $E(Y|A)$ as a function of the set A.

Lemma 5.3. *If B is the disjoint union of $A_1, \ldots, A_k$, then*

$$E(Y; B) = \sum_{j=1}^{k} E(Y; A_j)$$

Proof. Our assumption implies $Y1_B = \sum_{j=1}^{k} Y1_{A_j}$, so taking expected values, we have

$$E(Y; B) = E(Y1_B) = E\left(\sum_{j=1}^{k} Y1_{A_j}\right) = \sum_{j=1}^{k} E(Y1_{A_j}) = \sum_{j=1}^{k} E(Y; A_j) \qquad \square$$

Lemma 5.4. *If B is the disjoint union of $A_1, \ldots, A_k$, then*

$$E(Y|B) = \sum_{j=1}^{k} E(Y|A_j) \cdot \frac{P(A_j)}{P(B)}$$

In particular when $B = \Omega$ we have $EY = \sum_{j=1}^{k} E(Y|A_j) \cdot P(A_j)$.

Proof. Using the definition of conditional expectation, Lemma 5.3, then doing some arithmetic and using the definition again, we have

$$E(Y|B) = E(Y; B)/P(B) = \sum_{j=1}^{k} E(Y; A_j)/P(B)$$

$$= \sum_{j=1}^{k} \frac{E(Y; A_j)}{P(A_j)} \cdot \frac{P(A_j)}{P(B)} = \sum_{j=1}^{k} E(Y|A_j) \cdot \frac{P(A_j)}{P(B)}$$

which proves the desired result. $\qquad \square$

In the discussion in this section we have concentrated on the properties of conditional expectation given a single set A. To connect with more advanced treatments, we note that given a partition $\mathcal{A} = \{A_1, \ldots A_n\}$ of the sample space (i.e., disjoint sets whose union in Ω) then the conditional expectation given the partition is a random variable:

$$E(X|\mathcal{A}) = E(X|A_i) \quad \text{on } A_i$$

In this setting, Lemma 5.4 says

$$E[E(X|\mathcal{A})] = EX$$

i.e., the random variable $E(X|\mathcal{A})$ has the same expected value as X. Lemma 5.1 says that if X is constant on each part of the partition then

$$E(XY|\mathcal{A}) = XE(Y|\mathcal{A})$$

5.2   Examples

We begin by giving the definition of a martingale. Thinking of M_n as the amount of money at time n for a gambler betting on a fair game, and X_n as the outcomes of the gambling game we say that $M_0, M_1, \ldots$ is a **martingale** with respect to $X_0, X_1, \ldots$ if

(i) for any $n \geq 0$ we have $E|M_n| < \infty$

(ii) the value of M_n can be determined from the values for $X_n, \ldots X_0$ and M_0

and (iii) for any possible values $x_n, \ldots, x_0$

$$E(M_{n+1} - M_n|X_n = x_n, X_{n-1} = x_{n-1}, \ldots X_0 = x_0, M_0 = m_0) = 0 \qquad (5.4)$$

It will take several examples to explain why this is a useful definition.

- The first condition, $E|M_n| < \infty$, is needed to guarantee that the conditional expectation makes sense.
- To motivate the second, we note that in many of our examples X_n will be a Markov chain and $M_n = f(X_n, n)$. The conditioning event is formulated in terms of X_n because in passing from the random variables X_n that are driving the process the martingale to M_n, there may be a loss of information. For example, in Example 5.3 $M_n = X_n^2 - n$.
- The third, defining property, says that conditional on the past up to time n, the average profit from the bet on the nth game is 0.

To explain the reason for our interest in martingales, we will now give a number of examples. In what follows we will often be forced to write the conditioning event, so we introduce the short hand

$$A_v = \{X_n = x_n, X_{n-1} = x_{n-1}, \ldots, X_0 = x_0, M_0 = m_0\} \qquad (5.5)$$

where v is short for the vector $(x_n, \ldots, x_0, m_0)$.

Example 5.2 (Random Walks). Let $X_1, X_2, \ldots$ be i.i.d. with $EX_i = \mu$. Let $S_n = S_0 + X_1 + \cdots + X_n$ be a random walk. $M_n = S_n - n\mu$ is a martingale with respect to X_n.

Proof. To check this, note that $M_{n+1} - M_n = X_{n+1} - \mu$ is independent of $X_n, \dots, X_0, M_0$, so by (5.1) the conditional mean of the difference is just the mean:

$$E(M_{n+1} - M_n | A_v) = EX_{n+1} - \mu = 0 \qquad \square$$

In most cases, casino games are not fair but biased against the player. We say that M_n is a **supermartingale** with respect to X_n if a gambler's expected winnings on one play are negative:

$$E(M_{n+1} - M_n | A_v) \leq 0$$

To help remember the direction of the inequality, note that there is nothing "super" about a supermartingale. The definition traces its roots to the notion of superharmonic functions whose values at a point exceed the average value on balls centered around the point. If we reverse the sign and suppose

$$E(M_{n+1} - M_n | A_v) \geq 0$$

then M_n is called a **submartingale** with respect to X_n. A simple modification of the proof for Example 5.2 shows that if $\mu \leq 0$, then S_n defines a supermartingale, while if $\mu \geq 0$, then S_n is a submartingale.

Example 5.3 (Random Walks, II). Let $X_1, X_2, \dots$ be independent and identically distributed with $EX_i = 0$ and $\text{var}(X_i) = E(X_i^2) = \sigma^2$. Then $M_n = S_n^2 - n\sigma^2$ is a martingale with respect to X_n.

Proof. We begin with a little algebra:

$$M_{n+1} - M_n = (S_n + X_{n+1})^2 - S_n^2 - \sigma^2 = 2X_{n+1}S_n + X_{n+1}^2 - \sigma^2$$

Taking the conditional probability now and using (5.3) and (5.1)

$$E(2X_{n+1}S_n + X_{n+1}^2 - \sigma^2 | A_v) = 2S_n E(X_{n+1}|A_v) + E(X_{n+1}^2 - \sigma^2|A_v) = 0$$

since X_{n+1} is independent of A_v and has $EX_{n+1} = 0$ and $EX_{n+1}^2 = \sigma^2$. $\qquad \square$

Example 5.4 (Products of Independent Random Variables). To build a discrete time model of the stock market we let $X_1, X_2, \dots$ be independent ≥ 0 with $EX_i = 1$. Then $M_n = M_0 X_1 \cdots X_n$ is a martingale with respect to X_n. To prove this we note that (5.3)

$$E(M_{n+1} - M_n | A_v) = M_n E(X_{n+1} - 1|A_v) = 0$$

The reason for a multiplicative model is that changes in stock prices are thought to be proportional to its value. Also, in contrast to an additive model, we are guaranteed that prices will stay positive.

The last example generalizes easily to give:

Example 5.5 (Exponential Martingale). Let $Y_1, Y_2, \ldots$ be independent and identically distributed with $\phi(\theta) = E \exp(\theta Y_1) < \infty$. Let $S_n = S_0 + Y_1 + \cdots + Y_n$. Then $M_n = \exp(\theta S_n)/\phi(\theta)^n$ is a martingale with respect to Y_n. In particular, if $\phi(\theta) = 1$, then $\phi(\theta S_n)$ is a martingale.

Proof. If we let $X_i = \exp(\theta Y_i)/\phi(\theta)$, then $M_n = M_0 X_1 \cdots X_n$ with $EX_i = 1$ and this reduces to the previous example. $\qquad\square$

Example 5.6 (Gambler's Ruin). Let $X_1, X_2, \ldots$ be independent with

$$P(X_i = 1) = p \quad \text{and} \quad P(X_i = -1) = 1 - p$$

where $p \in (0, 1)$ and $p \neq 1/2$. Let $S_n = S_0 + X_1 + \cdots + X_n$. $M_n = \left(\frac{1-p}{p}\right)^{S_n}$ is a martingale with respect to X_n.

Proof. Pick θ so that $e^{\theta} = (1-p)/p$ and hence $e^{\theta} = p/(1-p)$. In this case

$$\phi(\theta) = Ee^{\theta X_i} = \frac{1-p}{p} \cdot p + \frac{p}{1-p} \cdot (1-p) = 1$$

so it follows from Example 5.5 that $\exp(\theta S_n) = ((1-p)/p)^{S-n}$ is a martingale. $\quad\square$

The next result will lead to a number of examples.

Theorem 5.1. *Let X_n be a Markov chain with transition probability p and let $f(x, n)$ be a function of the state x and the time n so that*

$$f(x, n) \geq \sum_y p(x, y)f(y, n + 1)$$

Then $M_n = f(X_n, n)$ is a supermartingale with respect to X_n.

If we have $\leq$ instead, then applying the result to $-f$ we see that $f(X_n, n)$ is a submartingale. If we have $=$, then using the results for $\leq$ and $\geq$ we see that $f(X_n, n)$ is a martingale.

Proof. By the Markov property and our assumption on f

$$E(f(X_{n+1}, n + 1)|A_v) \geq \sum_y p(x_n, y)f(y, n + 1) = f(x_n, n)$$

which proves the desired result. $\qquad\square$

The next two examples are useful for studying hitting probabilities and exit times.

Example 5.7. Consider a Markov chain X_n with finite state space S. Let $A \subset S$, and let $V_A = \min\{n \geq 0 : X_n \in A\}$. If for $x \in C = S - A$ we have

$$h(x) \geq \sum_y p(x, y)h(y)$$

then $h(X_{n \wedge V_A})$ is a supermartingale.

Proof. Let $\bar{X}_n$ be a modification of X_n in which all sites in A are absorbing and apply Theorem 5.1. □

Example 5.8. Continuing to use the assumptions of the last example, we assume that for $x \in C$ we have

$$g(x) \geq \alpha + \sum_y p(x, y)g(y)$$

Adding n to each side of the equation, and applying Theorem 5.1 to $f(x, n) = g(x) + \alpha n$ to $\bar{X}_n$ defined in the previous example, we see that $g(X_{V_A \wedge n}) + \alpha(V_A \wedge n)$ is a supermartingale.

5.3 Gambling Strategies, Stopping Times

The first result should be intuitive if we think of supermartingale as betting on an unfavorable game: the expected value of our fortune will decline over time.

Theorem 5.6. *If M_m is a supermartingale and $m \leq n$, then $EM_m \geq EM_n$.*

Proof. It is enough to show that the expected value decreases with each time step, i.e., $EM_k \geq EM_{k+1}$. To do this, we will again use the notation from (5.5)

$$A_v = \{X_n = x_n, X_{n-1} = x_{n-1}, \ldots, X_0 = x_0, M_0 = m\}$$

and note that linearity in the conditioning set (Lemma 5.3) and the definition of conditional expectation imply

$$E(M_{k+1} - M_k) = \sum_v E(M_{k+1} - M_k; A_v)$$

$$= \sum_v P(A_v)E(M_{k+1} - M_k | A_v) \leq 0$$

since supermartingales have $E(M_{k+1} - M_k | A_v) \leq 0$. □

As we argued at the end of the previous section, the result in Theorem 5.6 generalizes immediately to our other two types of processes. Multiplying by -1 we see

Theorem 5.7. *If M_m is a submartingale and $0 \le m < n$, then $EM_m \le EM_n$.*

Since a process is a martingale if and only if it is both a supermartingale and submartingale, we can conclude that:

Theorem 5.8. *If M_m is a martingale and $0 \le m < n$, then $EM_m = EM_n$.*

The most famous result of martingale theory (see Theorem 5.9) is that

$$\text{"you can't beat an unfavorable game."} \tag{5.6}$$

To lead up to this result, we will analyze a famous gambling system and show why it doesn't work.

Example 5.9 (Doubling Strategy). Suppose you are playing a game in which you will win or lose \$1 on each play. If you win you bet \$1 on the next play but if you lose, then you bet twice the previous amount. The idea behind the system can be seen by looking at what happens if we lose four times in a row and, then win:

outcome	L	L	L	L	W
bet	1	2	4	8	16
net profit	-1	-3	-7	-15	1

In this example our net profit when we win is \$1. Since $1 + 2 + \cdots + 2^k = 2^{k+1} - 1$, this is true if we lose k times in a row before we win. Thus every time we win our net profit is up by \$1 from the previous time we won.

This system will succeed in making us rich as long as the probability of winning is positive, so where's the catch? To explain, suppose that winning and losing each have probability 1/2, suppose that we start by betting \$1 and that we play the game 10 times. For example, our outcomes might be

$$W \quad L \quad W \quad W \quad L \quad L \quad W \quad L \quad L \quad L$$

Let J be the number of times we win $J = 4$ in the example above. By the way the doubling strategy is constructed, if we ignore the string of losses at the end, we have won \$$J$, which is binomial$(10, 1/2)$, so $EJ = 10/2 = 5$.

Let K be the number of the trial on which we last win with $K = 0$ if we never won In the example $K = 7$, and our losses are $-1 - 2 - 4- = -7$. A little thought shows

K	0	1	2	3	4	5	6	7	8	9	10
losses	1023	511	255	127	63	31	15	7	3	1	0
prob	$\frac{1}{1024}$	$\frac{1}{1024}$	$\frac{1}{512}$	$\frac{1}{256}$	$\frac{1}{128}$	$\frac{1}{64}$	$\frac{1}{32}$	$\frac{1}{16}$	$\frac{1}{8}$	$\frac{1}{4}$	$\frac{1}{2}$

Multiplying the loses by the probability and adding up the products we get

$$1 + 9 \cdot \frac{1}{2} - \frac{1}{4} - \frac{1}{8} \cdots - \frac{1}{512} - \frac{2}{1024} = 5$$

Thus the expected value of our losses at the end of our ten trials is equal to the expected amount that we have won, i.e., we have not changed the fact that it is a fair game. What the doubling strategy has done is to make the distribution like the reverse of a lottery ticket. About \$1 of our expected loss comes from the very unlikely event $K = 0$ while 511, 255, 127, and 63 combine for another \$2. Thus \$3 of the expected losses comes from events with a total probability of $1/64$.

To formulate and prove (5.6) we will introduce a family of betting strategies that generalize the doubling strategy. The amount of money we bet on the nth game, H_n, clearly, cannot depend on the outcome of that game, nor is it sensible to allow it to depend on the outcomes of games that will be played later. We say that H_n is an admissible gambling strategy or **predictable process** if for each n the value of H_n can be determined from $X_{n-1}, X_{n-2}, \ldots, X_0, M_0$.

To motivate the next definition, which we give for a general M_m think of H_m as the amount of stock we hold between time $m - 1$ and m, and M_m the price of a stock at time m. Then our wealth at time n is

$$W_n = W_0 + \sum_{m=1}^{n} H_m (M_m - M_{m-1}) \tag{5.7}$$

since the change in our wealth from time $m - 1$ to m is the amount we hold times the change in the price of the stock: $H_m (M_m - M_{m-1})$. To formulate the doubling strategy in this setting, let $X_m = 1$ if the result of the mth game is a win and -1 if the mth result is a loss, and let $M_n = X_1 + \cdots + X_n$ be the net profit of a gambler who bets one unit every time.

Theorem 5.9. *Suppose that M_n is a supermartingale with respect to X_n, H_n is predictable, and $0 \leq H_n \leq c_n$ where c_n is a constant that may depend on n. Then*

$$W_n = W_0 + \sum_{m=1}^{n} H_m (M_m - M_{m-1}) \quad \text{is a supermartingale}$$

We need the condition $H_n \geq 0$ to prevent the bettor from becoming the house by betting a negative amount of money. The upper bound $H_n \leq c_n$ is a technical condition that is needed to have expected values make sense. In the gambling context this assumption is harmless: even if the bettor wins every time there is an upper bound to the amount of money he can have at time n.

Proof. The change in our wealth from time n to time $n + 1$ is

$$W_{n+1} - W_n = H_{n+1}(Y_{n+1} - Y_n)$$

Continuing to use the notation introduced in (5.5)

$$A_v = \{X_n = x_n, X_{n-1} = x_{n-1}, \ldots, X_0 = x_0, M_0 = m_0\}.$$

we note that H_{n+1} is constant on the event A_v, so Lemma 5.1 implies

$$E(H_{n+1}(M_{n+1} - M_n)|A_v) = H_{n+1}E(M_{n+1} - M_n|A_v) \le 0$$

verifying that W_n is a supermartingale.   □

Arguing as in the discussion after Theorem 5.6 the same result holds for submartingales when $0 \le H_m \le c_n$ and for martingales with only the assumption that $|H_n| \le c_n$.

Though Theorem 5.9 may be depressing for gamblers, a simple special case gives us an important computational tool. To introduce this tool, we need one more notion. As in Sect. 1.3, we say that T is a **stopping time with respect to** X_n if the occurrence (or nonoccurrence) of the event $\{T = n\}$ can be determined from the information known at time n, $X_n, X_{n-1} \ldots X_0, M_0$.

Example 5.10 (Constant Betting Up to a Stopping Time). One possible gambling strategy is to bet \$1 each time until you stop playing at time T. In symbols, we let

$$H_m = \begin{cases} 1 & \text{if } T \ge m \\ 0 & \text{otherwise} \end{cases}$$

To check that this is an admissible gambling strategy we note that

$$\{H_m = 0\} = \{T \ge m\}^c = \{T \le m - 1\} = \cup_{k=1}^{m-1}\{T = k\}$$

By the definition of a stopping time, the event $\{T = k\}$ can be determined from the values of $M_0, X_0, \ldots, X_k$. Since the union is over $k \le m - 1$, H_m can be determined from the values of $M_0, X_0, X_1, \ldots, X_{m-1}$.

Having introduced the gambling strategy "Bet \$1 on each play up to time T" our next step is to compute the payoff we receive when $W_0 = M_0$. Letting $T \wedge n$ denote the minimum of T and n, i.e., it is T if $T < n$ and n if $T \ge n$, we can give the answer as:

$$W_n = M_0 + \sum_{m=1}^{n} H_m(M_m - M_{m-1}) = M_{T \wedge n} \tag{5.8}$$

To check the last equality, consider two cases:

(i) if $T \ge n$, then $H_m = 1$ for all $m \le n$, so

$$W_n = M_0 + (M_n - M_0) = M_n$$

(ii) if $T \le n$, then $H_m = 0$ for $m > T$, so the sum in (5.8) stops at T. In this case,

$$W_n = M_0 + (M_T - M_0) = M_T$$

The next picture should help with checking the last two results. The placement of the H_m is to remind us that the contribution of that bet to our wealth is $H_m(X_m - X_{m-1})$.

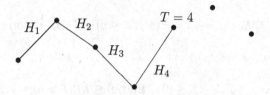

Combining (5.8) with Theorem 5.9 and using Theorem 5.6 we have

Theorem 5.10. *If M_n is a supermartingale with respect to X_n and T is a stopping time, then the stopped process $M_{T \wedge n}$ is a supermartingale with respect to X_n. In particular, $EM_{T \wedge n} \le EM_0$*

As in the discussion after Theorem 5.6, the analogous results are true for submartingales ($EM_{T \wedge n} \ge EM_0$) and for martingales ($EM_{T \wedge n} = EM_0$).

5.4 Applications

In this section we will apply the results from the previous section to rederive some of the results from Chap. 1 about hitting probabilities and exit times for random walks. To do this we will invent martingales M_n and stopping times T and then use the fact that $EM_T = EM_0$. Unfortunately, (i) all Theorem 5.10 guarantees is that $EM_{T \wedge n} = EM_0$ and (ii) a simple example shows that $EM_T = EM_0$ is not always true.

Example 5.11 (Bad Martingale). Let $X_1, X_2, \ldots X_n$ be independent with

$$P(X_i = 1) = P(X_i = -1) = 1/2$$

and define symmetric simple random walk by $S_n = S_0 + X_1 + \cdots + X_n$. Suppose $S_0 = 1$, let $V_0 = \min\{n : S_n = 0\}$ and $T = V_0$. From Sect. 1.11, we know that $P_1(T < \infty) = 1$. This implies $S_T = 0$ and hence $E_1 S_T = 0 \ne 1$. The trouble is that

$$P_1(V_N < V_0) = 1/N$$

so the random walk can visit some very large values before returning to 0. In particular if we let $R = \max_{0 \le m \le T} S_m$, then

$$ER = \sum_{m=1}^{\infty} P(R \ge m) = \sum_{m=1}^{\infty} P_1(V_m < V_0) = \sum_{m=1}^{\infty} 1/m = \infty.$$

The next result gives a simple condition that rules out the problem in the last example. In more advanced treatments it is called the bounded convergence theorem.

Theorem 5.11. *Suppose M_n is a martingale and T a stopping time with $P(T < \infty)=1$ and $|M_{T\wedge n}| \leq K$ for some constant K. Then $EM_T = EM_0$.*

Proof. Theorem 5.10 implies

$$EM_0 = EM_{T\wedge n} = E(M_T; T \leq n) + E(M_n; T > n).$$

The second term $\leq KP(T > n)$ and

$$|E(M_T; T \leq n) - E(M_T)| \leq KP(T > n)$$

Since $P(T > n) \to 0$ as $n \to \infty$ the desired result follows. $\square$

5.4.1 Exit Distributions

In Example 5.7 we considered a Markov chain with state space S and $A, B \subset S$, so that $C = S - (A \cup B)$ is finite. Let $V_A = \min\{n \geq 0 : X_n \in A\}$ and suppose that $P_x(V_A \wedge V_B < \infty) > 0$ for all $x \in C$. We showed in Theorem 1.28 that if $h(a) = 1$ for $a \in A$, $h(b) = 0$ for $b \in B$, and for $x \in C$ we have

$$h(x) = \sum_y p(x, y)h(y)$$

then $h(x) = P_x(V_A < V_B)$.

Proof. As before it follows from Lemma 1.3 that $P_x(V_A \wedge V_B < \infty) = 1$ for all $x \in C$. Let $T = V_A \wedge V_B$. Lemma 5.7 implies that $h(X_{n\wedge T})$ is a martingale. Since C is finite, $h = 1$ on A and $h = 0$ on B, there is a K so that $h(X_{n\wedge T}) \leq K$ and the desired result follows from Theorem 5.11. $\square$

Example 5.12 (Exit Distribution for Simple Random Walk). Let $M_n = S_n$ and $T = \min\{n : S_n \notin (a, b)\}$. $M_{T\wedge n} \in [a, b]$ so it is bounded. Thus using Theorem 5.11 we can conclude

$$x = E_x S_T = aP_x(S_T = a) + b(1 - P(S_T = a)),$$

and then solve to conclude

$$P_x(S_T = a) = \frac{b - x}{b - a} \qquad P_x(S_T = b) = \frac{x - a}{b - a} \tag{5.9}$$

Turning now to the asymmetric case.

Example 5.13 (Gambler's Ruin). Let $S_n = S_0 + X_1 + \cdots X_n$ where $X_1, X_2, \ldots X_n$ are independent with

$$P(X_i = 1) = p \quad \text{and} \quad P(X_i = -1) = q = 1 - p.$$

Suppose $p \in (0,1)$, $p \neq 1/2$ and let $h(x) = (q/p)^x$. Example 5.6 implies that $M_n = h(S_n)$ is a martingale. Let $T = \min\{n : S_n \notin (a,b)\}$. It is easy to see that T is a stopping time. Lemma 1.3 implies that $P(T < \infty) = 1$.

$$M_{T \wedge n} \leq \max\{((1-p)/p)^a, ((1-p)/p)^a\},$$

so using Theorem 5.11.

$$(q/p)^x = E_x(q/x)^{S(\tau)} = (q/p)^a P(S_\tau = a) + (q/p)^b[1 - P(S_\tau = a)] \qquad (5.10)$$

Solving gives

$$P_x(S_\tau = a) = \frac{(q/p)^b - (q/p)^x}{(q/p)^b - (q/p)^a} \qquad (5.11)$$

generalizing (1.24).

Example 5.14 (General Birth and Death Chains). The state space is $\{0,1,2,\ldots\}$ and the transition probability has

$$
\begin{aligned}
p(x, x+1) &= p_x \\
p(x, x-1) &= q_x & \text{for } x > 0 \\
p(x, x) &= 1 - p_x - q_x & \text{for } x \geq 0
\end{aligned}
$$

while the other $p(x,y) = 0$. Let $V_y = \min\{n \geq 0 : X_n = y\}$ be the time of the first visit to y and let $\tau = V_a \wedge V_b$. Let $\phi(z) = \sum_{y=1}^{z} \prod_{x=1}^{y-1} q_x/p_x$. The definition of ϕ implies $\phi(x+1) - \phi(x) = (q_x/p_x)(\phi(x) - \phi(x-1))$, so $E_x\phi(X_1) = \phi(x)$ for $a < x < b$, and it follows from Example 5.7 that $\phi(X_{\tau \wedge n})$ is a martingale. Using Theorem 5.11 now we have

$$\phi(x) = E_x\phi(X_\tau) = \phi(a)P_x(X_\tau = a) + \phi(b)[1 - P_x(X_\tau = a)]$$

and a little algebra gives

$$P_x(V_a < V_b) = \frac{\phi(x) - \phi(a)}{\phi(b) - \phi(a)}$$

From this it follows that 0 is recurrent if and only if $\phi(b) \to \infty$ as $b \to \infty$, giving another solution of Exercise 1.70 from Chap. 1.

5.4.2 Exit Times

Consider a Markov chain with state space S and let $A \subset S$ so that $C = S - A$ is finite. Suppose that $P_x(V_A < \infty) > 0$ for each $x \in C$. In Theorem 1.29, we showed that if $g(x) = 0$ for $x \in A$ and

$$g(x) = 1 + \sum_y p(x, y) g(y) \qquad \text{for } x \in C$$

then $g(x) = E_x V_A$.

Proof. As before it follows from Lemma 1.3 that $P_x(V_A < \infty) = 1$ for all $x \in C$. Lemma 5.8 implies that $g(X_{V_A \wedge n}) + (V_A \wedge n)$ is a martingale. Taking expected value we have

$$g(x) = E_x g(X_{V_A \wedge n}) + E_x(V_A \wedge n)$$

Since g is bounded, $E_x g(X_{V_A \wedge n}) \to 0$. To handle the second term we note that by (1.6)

$$E_x(V_A \wedge n) = \sum_{m=0}^{n} P(V_A \geq m) \uparrow \sum_{m=0}^{\infty} P(V_A \geq m) = E_x V_A. \qquad (5.12)$$

which gives the desired result.   □

The reasoning in (5.12) is sometimes called the monotone convergence theorem. See Theorem A.11.

Example 5.15 (Duration of Fair Games). Let S_n be symmetric simple random walk. Let $T = \min\{n : S_n \notin (a, b)\}$ where $a < 0 < b$. Our goal here is to prove a generalization of (1.29):

$$E_x T = (x - a)(b - x)$$

It suffices to show $E_0 T = -ab$. Example 5.3 implies that $S_n^2 - n$ is a martingale. If we argue casually $0 = E_0(S_T^2 - T)$ so using (5.9)

$$E_0(T) = E_0(S_T^2) = a^2 P_0(S_T = a) + b^2 P_0(S_T = b)$$

$$= a^2 \frac{b}{b - a} + b^2 \frac{-a}{b - a} = ab \frac{a - b}{b - a} = -ab$$

However, this is time $S_{T \wedge n}^2 - (T \wedge n)$ is not bounded so we can't use Theorem 5.11. To give a rigorous proof now, we use Theorem 5.10 to conclude

$$0 = E_0(S^2_{T \wedge n} - T \wedge n) = a^2 P(S_T = a, \tau \le n) + b^2 P(S_T = b, T \le n)$$
$$+ E(S^2_n; T > n) - E_0(T \wedge n)$$

$P(T < \infty) = 1$ and $S^2_{T \wedge n} \le \max\{a^2, b^2\}$ so the third term tends to 0. Using (5.12) $E_0(T \wedge n) \to E_0 T$. Putting it all together, we have

$$0 = a^2 P_0(S_\tau = a) + b^2 P_0(S_\tau = b) - E_0 \tau$$

and we have proved the result.

Consider now a random walk $S_n = S_0 + X_1 + \cdots + X_n$ where $X_1, X_2, \ldots$ are i.i.d. with mean μ and S_0 is not random. From Example 5.2, $M_n = S_n - n\mu$ is a martingale with respect to X_n.

Theorem 5.12 (Wald's Equation). *If T is a stopping time with $ET < \infty$, then*

$$E(S_T - S_0) = \mu ET$$

Recalling Example 5.11, which has $\mu = 0$ and $S_0 = 1$, but $S_T = 1$ shows that for symmetric simple random walk $E_1 V_0 = \infty$.

Why is this true? Since $M_n = S_n - n\mu$ is a martingale using $EM_T = EM_0$ gives $ES_0 = ES_T - \mu ET$.

Proof. To make the last calculation rigorous, we need to stop at time $T \wedge n$ and let $n \to \infty$. Theorem 5.10 gives

$$ES_0 = E(S_{T \wedge n}) - \mu E(T \wedge n)$$

As $n \uparrow \infty$, $E_0(T \wedge n) \uparrow E_0 T$ by (5.12). To pass to the limit in the other term, we note that

$$|S_{T \wedge n}| \le |S_0| + Y \quad \text{where} \quad Y = \sum_{m=1}^{\infty} |X_m| 1_{\{T \ge m\}}$$

Since $\{T \ge m\} = \{T \le m - 1\}^c$ is determined by $S_0, X_1, \ldots X_{m-1}$, it is independent of X_m, and

$$EY = |EX_i| \sum_{m=1}^{\infty} P(T \ge m) = |EX_i| ET < \infty$$

Using the dominated convergence theorem, Theorem A.12, now we see that $ES_{T \wedge n} \to ES_T$ and the desired result follows. $\qquad\square$

Example 5.16 (Duration of Unfair Games). Consider asymmetric simple random walk $S_n = S_0 + X_1 + \cdots + X_n$ where $X_1, X_2, \ldots X_n$ be independent with

$$P(X_i = 1) = p \quad \text{and} \quad P(X_i = -1) = q = 1 - p$$

Suppose $0 < p < 1/2$, and let $V_0 = \min\{n \geq 0 : S_n = 0\}$. Noting that $EX_i = p - q$ and using Theorem 5.10 we have

$$x = E_x[S(V_0 \wedge n) - (p - q)(V_0 \wedge n)] \geq (q - p)E_x(V_0 \wedge n)$$

Letting $n \to \infty$ we see that $E_x(V_0) \leq x/(q-p) < \infty$ so we can use Wald's equation to conclude that

$$x = E_x[S(V_0 \wedge n) - (p - q)(V_0 \wedge n)]$$

and hence

$$E_x V_0 = x/(q - p). \tag{5.13}$$

5.4.3   Extinction and Ruin Probabilities

In this subsection we study the probability a process reaches 0 or $(-\infty, 0]$. Our next two examples use the exponential martingale in Example 5.5.

Example 5.17 (Left-Continuous Random Walk). Suppose that $X_1, X_2, \ldots$ are independent integer valued random variables with $EX_i > 0$, $P(X_i \geq -1) = 1$, and $P(X_i = -1) > 0$. These walks are called left-continuous since they cannot jump over any integers when they are decreasing, which is going to the left as the number line is usually drawn. Let $\phi(\theta) = \exp(\theta X_i)$ and define $\alpha < 0$ by the requirement that $\phi(\alpha) = 1$. To see that such an α exists, note that (i) $\phi(0) = 1$ and

$$\phi'(\theta) = \frac{d}{d\theta} E e^{\theta x_i} = E(x_i e^{\theta x_i}) \quad \text{so} \quad \phi'(0) = Ex_i > 0$$

and it follows that $\phi(\theta) < 1$ for small negative θ. (ii) If $\theta < 0$, then $\phi(\theta) \geq e^{-\theta} P(x_i = -1) \to \infty$ as $\theta \to -\infty$. Our choice of α makes $\exp(\alpha S_n)$ a martingale. Having found the martingale it is easy now to conclude:

Theorem 5.13. *Consider a left-continuous random walk with positive mean. Let $x > 0$ and $V_0 = \min\{n : S_n = 0\}$.*

$$P_x(V_0 < \infty) = e^{\alpha x}$$

Proof. Again if one argues casually

$$e^{\alpha x} = E_x(\exp(\alpha V_0)) = P_x(V_0 < \infty)$$

but we have to prove that there is no contribution from $\{V_0 = \infty\}$. To do this note that Theorem 5.10 gives

$$e^{\alpha x} = E_x \exp(\alpha S_{V_a \wedge n}) = P_x(V_a \leq n) + E_x(\exp(\alpha S_n); V_0 > n)$$

$\exp(\alpha S_n) \leq 1$ on $V_0 > n$ but since $P(V_0 = \infty) > 0$ this is not enough to make the last term vanish. The strong law of large numbers implies that on $V_0 = \infty$, $S_n/n \to \mu > 0$, so the second term $\to 0$ as $n \to \infty$ and it follows that $e^{\alpha x} = e^{\alpha a} P_0(V_a < \infty)$.   □

Example 5.18 (Extinction Probability for Branching Process). Consider for sim-plicity a continuous time branching process in which individuals live for an exponentially distributed amount of time and then die leaving k offspring with probability p_k. Let $Y_1, Y_2, \ldots$ be independent with $P(Y_i = k) = p_k$. While there are a positive number of individuals the branching process has embedded jump chain

$$Z_{n+1} = Z_n + Y_{n+1} - 1$$

To have a random walk, we will use this definition for all n. Since $Y_{n+1} \geq 0$ this is a left-continuous random walk with jumps $X_k = Y_k - 1$. If $EY_k > 1$, then $EX_k > 0$ so we can use Theorem 5.13 to conclude that if $\alpha < 0$ has $Ee^{\alpha X_k} = 1$, then $P_1(V_0 < \infty) = e^{\alpha}$. Since $Y_k = X_k + 1$ it follows that $Ee^{\alpha Y_k} = e^{\alpha}$. If we write $\rho = e^{\alpha}$ and define the generating function by

$$\psi(z) = \sum_{k=0}^{\infty} z^k p_k z^k = E(z^{Y-k})$$

what we have shown is that $\psi(\rho) = \rho$.

When the random walk is not left-continuous we cannot get exact results on hitting probabilities but we can still get a bound.

Example 5.19 (Cramér's Estimate of Ruin). Let S_n be the total assets of an insur-ance company at the end of year n. During year n, premiums totaling c dollars are received, while claims totaling Y_n dollars are paid, so

$$S_n = S_{n-1} + c - Y_n$$

Let $X_n = c - Y_n$ and suppose that $X_1, X_2, \ldots$ are independent random variables that are normal with mean $\mu > 0$ and variance σ^2. That is the density function of X_i is

$$(2\pi\sigma^2)^{-1/2} \exp(-(x-\mu)^2/2\sigma^2)$$

Let B for bankrupt be the event that the wealth of the insurance company is negative at some time n. We will show

$$P(B) \le \exp(-2\mu S_0/\sigma^2) \tag{5.14}$$

In words, in order to be successful with high probability, $\mu S_0/\sigma^2$ must be large, but the failure probability decreases exponentially fast as this quantity increases.

Proof. We begin by computing $\phi(\theta) = E\exp(\theta X_i)$. To do this we need a little algebra

$$-\frac{(x-\mu)^2}{2\sigma^2} + \theta(x-\mu) + \theta\mu = -\frac{(x-\mu-\sigma^2\theta)^2}{2\sigma^2} + \frac{\sigma^2\theta^2}{2} + \theta\mu$$

and a little calculus

$$\phi(\theta) = \int e^{\theta x}(2\pi\sigma^2)^{-1/2}\exp(-(x-\mu)^2/2\sigma^2)\,dx$$

$$= \exp(\sigma^2\theta^2/2 + \theta\mu)\int (2\pi\sigma^2)^{-1/2}\exp\left(-\frac{(x-\mu-\sigma^2\theta)^2}{2\sigma^2}\right)dx$$

Since the integrand is the density of a normal with mean $\mu + \sigma^2\theta$ and variance σ^2 it follows that

$$\phi(\theta) = \exp(\sigma^2\theta^2/2 + \theta\mu) \tag{5.15}$$

If we pick $\theta = -2\mu/\sigma^2$, then

$$\sigma^2\theta^2/2 + \theta\mu = 2\mu^2/\sigma^2 - 2\mu^2/\sigma^2 = 0$$

So Example 5.5 implies $\exp(-2\mu S_n/\sigma^2)$ is a martingale. Let $T = \min\{n : S_n \le 0\}$. Theorems 5.10 and 5.8 give

$$\exp(-2\mu S_0/\sigma^2) = E\exp(-2\mu S_{T\wedge n}) \ge P(T \le n)$$

since $\exp(-2\mu S_T/\sigma^2) \ge 1$ and the contribution to the expected value from $\{T > n\}$ is ≥ 0. Letting $n \to \infty$ now and noticing $P(T \le n) \to P(B)$ gives the desired result. $\square$

5.4.4 Positive Recurrence of the GI/G/1 Queue*

Let $t_1, t_2, \ldots$ be the interarrival times which are independent with distribution F and let $s_1, s_2, \ldots$ be the service times which are independent with distribution G. Let X_n

be the workload in the queue at the time of arrival of the nth customer, not counting the service time of the nth customer, s_n. X_n is a Markov chain with

$$X_{n+1} = (X_n + s_n - t_{n+1})^+$$

where $y^+ = \max\{y, 0\}$. To see this, note that the amount of work in front of the $(n+1)$th customer is that in front of the nth customer plus his service time s_n, minus the time t_{n+1} between the arrival of customers n and $n+1$. We take the positive part because if $X_n + s_n - t_{n+1} < 0$ the server has caught up and the workload in front of the $(n+1)$th customer is 0.

Our goal is to show

Theorem 5.14. *If $Et_i > Es_i$, then X_n is positive recurrent.*

The arguments were a little more sophisticated than others in the book but they constitute a powerful method for proving that processes are positive recurrent.

Lemma 5.15. *Let $\epsilon = (Es_0 - Et_1)/2$. There is a K so that $E_x(X_1 - x) \leq -\epsilon$ for $x > K$.*

Proof. To do this we note that

$$E(x + s_0 - t_1)^+ - x = E(x + s_0 - t_1; s_0 - t_1 \geq -x) - x$$

$$= E(s_0 - t_1; s_0 - t_1 \geq -x) - xP(s_0 - t_1 < -x)$$

$$\leq E(s_0 - t_1; s_0 - t_1 \geq -x) \to -2\epsilon$$

as $x \to \infty$, so if $x > K$, then $E(x + s_0 - t_1)^+ - x \leq -\epsilon$.   □

For the next two steps we treat a more general situation. In the case of the $GI/G/1$ queue $\phi(x) = x$.

Lemma 5.16. *Consider a Markov chain on $[0, \infty)$. Suppose that there is a $\phi(x) \geq 0$ with $\phi(x) \to \infty$ as $x \to \infty$, and a K so that if $x > K$, then $E_x\phi(X_1) \leq \phi(x)$. Let $U_K = \min\{n : X_n \leq K\}$. Then $E_xU_K \leq x/\epsilon$.*

Proof. It follows from Example 5.8 that $\phi(X_{U_K \wedge n}) + \epsilon(U_K \wedge n)$ is a supermartingale. Stopping at time $U_K \wedge n$ we have

$$\phi(x) \geq E_x\phi(X(U_K \wedge n)) + \epsilon E(U_k \wedge n) \geq \epsilon E(U_k \wedge n)$$

since $\phi \geq 0$. Letting $n \to \infty$ we have shown $E_xU_K \leq x/\epsilon$.   □

The final, somewhat messy, step is to show that $E_xV_0 < \infty$. To carry out the proof we assume (i) X_n is an irreducible chain on a countable state space or (ii) that X_n is the workload of a $GI/G1/1$ queue. These assumptions are needed so that the quantities l and ρ defined in the proof have $L < \infty$ and $\rho > 0$.

Lemma 5.17. *Suppose in addition to the assumptions in Lemma 5.16 that* $M = \sup\{E_x\phi(X_1) : x \leq K\} < \infty$. *There are constants* C_1 *and* C_2 *so that*

$$E_x V_0 \leq \frac{\phi(x) + C_1}{\epsilon} + C_2$$

Proof. Let $K(x) = \min\{k : p^k(x, 0) > 0\}, L = \max_{x \leq K} K(x)$, and $\rho = \min\{P_x(V_0 \leq L) : x \leq K\}$. Let $R_0 = \min\{n : X_n \leq K\}$. Let

$$S_1 = \min\{n > R_0 : X_n = 0, X_n > K, \text{ or } n \geq R_0 + L\}$$

If $X(S_1) = 0$, then we set $\bar{S}_1 = \infty$, since we have achieved our goal of reaching 0. If not, then $\bar{S}_1 = S_1$ and

$$R_1 = \min\{n \geq \bar{S}_1 : X_n \leq K\}$$
$$S_2 = \min\{n > R_1 : X_n = 0 \text{ or } n \geq R_1 + L\}$$

Let $N = \min\{n : \bar{S}_n = \infty\}$. By the definition of ρ, $P(N > k) \leq (1 - \rho)^k$ and hence $EN \leq 1/\rho$. Since we stop at $R_{k-1} + L$, we have $S_k - R_{k-1} \leq L$. By the definition of M and Lemma 5.16, if $\bar{S}_k < \infty$, then we have

$$E(R_k - \bar{S}_k) \leq M/\epsilon.$$

Note that $R_k - \bar{S}_k = 0$ when $X(\bar{S}_k) \leq K$.

We reach 0 at time S_N. Using the fact that $EN \leq 1/\rho$ with the bounds on $S_k - R_{k-1}$ and $E(R_k - S_k)$ the result now follows with $C_1 = M/\rho$ and $C_2 = L/\rho$.   □

5.5 Exercises

Throughout the exercises we will use our standard notion for hitting times. $T_a = \min\{n \geq 1 : X_n = a\}$ and $V_a = \min\{n \geq 0 : X_n = a\}$.

5.1. *Brother–sister mating.* Consider the six state chain defined in Exercise 1.66. Show that the total number of A's is a martingale and use this to compute the probability of getting absorbed into the 2,2 (i.e., all A's state) starting from each initial state.

5.2. *Lognormal stock prices.* Consider the special case of Example 5.4 in which $X_i = e^{\eta_i}$ where $\eta_i = \text{normal}(\mu, \sigma^2)$. For what values of μ and σ is $M_n = M_0 \cdot X_1 \cdots X_n$ a martingale?

5.3. Let X_n be the *Wright–Fisher model with no mutation* defined in Example 1.9. (a) Show that X_n is a martingale and use Theorem 5.11 to conclude that $P_x(V_N < V_0) = x/N$. (b) Show that $Y_n = X_n(N - X_n)/(1 - 1/N)^n$ is a martingale. (c) Use this to conclude that

$$(N - 1) \leq \frac{x(N - x)(1 - 1/N)^n}{P_x(0 < X_n < N)} \leq \frac{N^2}{4}$$

5.4. *An unfair fair game.* Define random variables recursively by $Y_0 = 1$ and for $n \geq 1$, Y_n is chosen uniformly on $(0, Y_{n-1})$. If we let $U_1, U_2, \ldots$ be uniform on $(0, 1)$, then we can write this sequence as $Y_n = U_n U_{n-1} \cdots U_0$. (a) Use Example 5.4 to conclude that $M_n = 2^n Y_n$ is a martingale. (b) Use the fact that $\log Y_n = \log U_1 + \cdots + \log U_n$ to show that $(1/n) \log X_n \to -1$. (c) Use (b) to conclude $M_n \to 0$, i.e., in this "fair" game our fortune always converges to 0 as time tends to ∞.

5.5. Consider a favorable game in which the payoffs are -1, 1, or 2 with probability $1/3$ each. Use the results of Example 5.17 to compute the probability we ever go broke (i.e., our winnings W_n reach \$0) when we start with \$$i$.

5.6. Let $S_n = X_1 + \cdots + X_n$ where the X_i are independent with $EX_i = 0$ and var $(X_i) = \sigma^2$. By example 5.3 $S_n^2 - n\sigma^2$ is a martingale. Let $T = \min\{n : |S_n| > a\}$. Use Theorem 5.10 to show that $ET \geq a^2/\sigma^2$. For simple random walk $\sigma^2 = 1$ and we have equality.

5.7. *Wald's second equation.* Let $S_n = X_1 + \cdots + X_n$ where the X_i are independent with $EX_i = 0$ and var $(X_i) = \sigma^2$. Use the martingale from the previous problem to show that if T is a stopping time with $ET < \infty$, then $ES_T^2 = \sigma^2 ET$.

5.8. *Variance of the time of gambler's ruin.* Let $X_1, X_2, \ldots$ be independent with $P(X_i = 1) = p$ and $P(X_i = -1) = q = 1 - p$ where $p < 1/2$. Let $S_n = S_0 + \xi_1 + \cdots + \xi_n$ and let $V_0 = \min\{n \geq 0 : S_n = 0\}$. (5.13) implies $E_x V_0 = x/(1 - 2p)$. The aim of this problem is to compute the variance of V_0. If we let $Y_i = X - i - (p - q)$ and note that $EY_i = 0$ and

$$\text{var}(Y_i) = \text{var}(X_i) = EX_i u^2 - (EX_i)^2$$

then it follows that $(S_n - (p - q)n)^2 - n(1 - (p - q)^2)$ is a martingale. (a) Use this to conclude that when $S_0 = x$ the variance of V_0 is

$$x \cdot \frac{1 - (p - q)^2}{(q - p)^3}$$

(b) Why must the answer in (a) be of the form cx?

5.9. *Generating function of the time of gambler's ruin.* Continue with the set-up of the previous problem. (a) Use the exponential martingale and our stopping theorem to conclude that if $\theta \leq 0$, then $e^{\theta x} = E_x(\phi(\theta)^{-V_0})$. (b) Let $0 < s < 1$. Solve the equation $\phi(\theta) = 1/s$, then use (a) to conclude

$$E_x(s^{V_0}) = \left(\frac{1 - \sqrt{1 - 4pqs^2}}{2ps} \right)^x$$

(c) Why must the answer in (b) be of the form $f(s)^x$?

5.10. Let Z_n be a branching process with offspring distribution p_k with $p_0 > 0$ and $\mu = \sum_k kp_k > 1$. Let $\phi(\theta) = \sum_{k=0}^{\infty} p_k \theta^k$. (a) Show that $E(\theta^{Z_{n+1}}|Z_n) = \phi(\theta)^{Z_n}$. (b) Let ρ be the solution < 1 of $\phi(\rho) = \rho$ and conclude that $P_k(T_0 < \infty) = \rho^k$

Chapter 6
Mathematical Finance

6.1 Two Simple Examples

To warm up for the developments in the next section we will look at two simple concrete examples under the unrealistic assumption that the interest rate is 0.

One Period Case In our first scenario the stock is at 90 at time 0 and may be 80 or 120 at time 1.

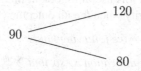

Suppose now that you are offered a **European call option** with **strike price** 100 and **expiry** 1. This means that after you see what happened to the stock, you have an option to buy the stock (but not an obligation to do so) for 100 at time 1. If the stock price is 80, you will not exercise the option to purchase the stock and your profit will be 0. If the stock price is 120 you will choose to buy the stock at 100 and then immediately sell it at 120 to get a profit of 20. Combining the two cases we can write the payoff in general as $(X_1 - 100)^+$, where $z^+ = \max\{z, 0\}$ denotes the positive part of z.

Our problem is to figure out the right price for this option. At first glance this may seem impossible since we have not assigned probabilities to the various events. However, it is a miracle of **"pricing by the absence of arbitrage"** that in this case we do not have to assign probabilities to the events to compute the price. To explain this we start by noting that X_1 will be 120 ("up") or 80 ("down") for a profit of 30 or a loss of 10, respectively. If we pay c for the option, then when X_1 is up we make a profit of $20 - c$, but when it is down we make $-c$. The last two sentences are summarized in the following table:

© Springer International Publishing Switzerland 2016
R. Durrett, *Essentials of Stochastic Processes*, Springer Texts in Statistics,
DOI 10.1007/978-3-319-45614-0_6

	stock	option
up	30	$20 - c$
down	-10	$-c$

Suppose we buy x units of the stock and y units of the option, where negative numbers indicate that we sold instead of bought. One possible strategy is to choose x and y so that the outcome is the same if the stock goes up or down:

$$30x + (20 - c)y = -10x + (-c)y$$

Solving, we have $40x + 20y = 0$ or $y = -2x$. Plugging this choice of y into the last equation shows that our profit will be $(-10 + 2c)x$. If $c > 5$, then we can make a large profit with no risk by buying large amounts of the stock and selling twice as many options. Of course, if $c < 5$, we can make a large profit by doing the reverse. Thus, in this case the only sensible price for the option is 5.

A scheme that makes money without any possibility of a loss is called an **arbitrage opportunity**. It is reasonable to think that these will not exist in financial markets (or at least be short-lived) since if and when they exist people take advantage of them and the opportunity goes away. Using our new terminology we can say that the only price for the option that is consistent with absence of arbitrage is $c = 5$, so that must be the price of the option.

To find prices in general, it is useful to look at things in a different way. Let $a_{i,j}$ be the profit for the ith security when the jth outcome occurs.

Theorem 6.1. *Exactly one of the following holds:*

(i) *There is an investment allocation x_i so that $\sum_{i=1}^{m} x_i a_{i,j} \geq 0$ for each j and $\sum_{i=1}^{m} x_i a_{i,k} > 0$ for some k.*
(ii) *There is a probability vector $p_j > 0$ so that $\sum_{j=1}^{n} a_{i,j} p_j = 0$ for all i.*

Here an x satisfying (i) is an arbitrage opportunity. We never lose any money but for at least one outcome we gain a positive amount. Turning to (ii), the vector p_j is called a martingale measure since if the probability of the jth outcome is p_j, then the expected change in the price of the ith stock is equal to 0. Combining the two interpretations we can restate Theorem 6.2 as:

Theorem 6.2. *There is no arbitrage if and only if there is a strictly positive probability vector so that all the stock prices are martingale.*

Proof. One direction is easy. If (i) is true, then for any strictly positive probability vector $\sum_{i=1}^{m} \sum_{j=1}^{n} x_i a_{i,j} p_j > 0$, so (ii) is false.

Suppose now that (i) is false. The linear combinations $\sum_{i=1}^{m} x_i a_{i,j}$ when viewed as vectors indexed by j form a linear subspace of n-dimensional Euclidean space. Call it $\mathcal{L}$. If (i) is false, this subspace intersects the positive orthant $\mathcal{O} = \{y : y_j \geq 0 \text{ for all } j\}$ only at the origin. By linear algebra we know that $\mathcal{L}$ can be extended to an $n - 1$ dimensional subspace $\mathcal{H}$ that only intersects $\mathcal{O}$ at the origin. (Repeatedly find a line not in the subspace that only intersects $\mathcal{O}$ at the origin and add it to the subspace.)

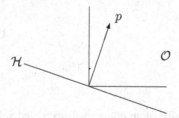

Since $\mathcal{H}$ has dimension $n - 1$, it can be written as $\mathcal{H} = \{y : \sum_{j=1}^{n} y_j p_j = 0\}$. By replacing p by $-p$ if necessary we can suppose p is a vector with at least one positive component. Since for each fixed i the vector $a_{i,j}$ is in $\mathcal{L} \subset \mathcal{H}$, $\sum_j a_{i,j} p_j = 0$. To see that all the $p_j > 0$ we leave it to the reader to check that if not, there would be a nonzero vector in $\mathcal{O}$ that would be in $\mathcal{H}$.   □

To apply Theorem 6.2 to our simplified example, we begin by noting that in this case $a_{i,j}$ is given by

		$j = 1$ up	$j = 2$ down
stock	$i = 1$	30	-10
option	$i = 2$	$20 - c$	$-c$

By Theorem 6.2 if there is no arbitrage, then there must be an assignment of probabilities p_j so that

$$30p_1 - 10p_2 = 0 \qquad (20 - c)p_1 + (-c)p_2 = 0$$

From the first equation we conclude that $p_1 = 1/4$ and $p_2 = 3/4$. Rewriting the second we have

$$c = 20p_1 = 20 \cdot (1/4) = 5$$

To prepare for the general case note that the equation $30p_1 - 10p_2 = 0$ says that under p_j the stock price is a martingale (i.e., the average value of the change in price is 0), while $c = 20p_1 + 0p_2$ says that the price of the option is then the expected value under the martingale probabilities.

Two-Period Binary Tree Suppose that a stock price starts at 100 at time 0. At time 1 (which we think of as one month later) it will either be worth 120 or 90. If the stock is worth 120 at time 1, then it might be worth 140 or 115 at time 2. If the price is 90 at time 1, then the possibilities at time 2 are 120 and 80. The last three sentences can be simply summarized by the following tree:

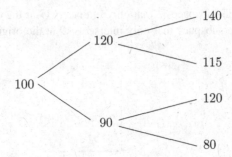

Using the idea that the value of an option is its expected value under the probability that makes the stock price a martingale, we can quickly complete the computations in our example. When $X_1 = 120$ the two possible scenarios lead to a change of $+20$ or -5, so the probabilities of these two events should be 1/5 and 4/5. When $X_1 = 90$ the two possible scenarios lead to a change of $+30$ or -10, so the probabilities of these two events should be 1/4 and 3/4. When $X_0 = 0$ the possible price changes on the first step are $+20$ and -10, so their probabilities are 1/3 and 2/3. Making a table of the possibilities, we have

X_1	X_2	probability	$(X_2 - 100)^+$
120	140	$(1/3) \cdot (1/5)$	40
120	115	$(1/3) \cdot (4/5)$	15
90	120	$(2/3) \cdot (1/4)$	20
90	80	$(2/3) \cdot (3/4)$	0

so the value of the option is

$$\frac{1}{15} \cdot 40 + \frac{4}{15} \cdot 15 + \frac{1}{6} \cdot 20 = \frac{80 + 120 + 100}{30} = 10$$

The last derivation may seem a little devious, so we will now give a second derivation of the price of the option based on absence of arbitrage. In the scenario described above, our investor has four possible actions:

A_0. Put \$1 in the bank and end up with \$1 in all possible scenarios.
A_1. Buy one share of stock at time 0 and sell it at time 1.
A_2. Buy one share at time 1 if the stock is at 120 and sell it at time 2.
A_3. Buy one share at time 1 if the stock is at 90 and sell it at time 2.

These actions produce the following payoffs in the indicated outcomes:

X_1	X_2	A_0	A_1	A_2	A_3	option
120	140	1	20	20	0	40
120	115	1	20	-5	0	15
90	120	1	-10	0	30	20
90	80	1	-10	0	-10	0

Noting that the payoffs from the four actions are themselves vectors in four dimensional space, it is natural to think that by using a linear combination of these actions we can reproduce the option exactly. To find the coefficients z_i for the actions A_i we write four equations in four unknowns,

$$z_0 + 20z_1 + 20z_2 = 40$$

$$z_0 + 20z_1 - 5z_2 = 15$$

$$z_0 - 10z_1 + 30z_3 = 20$$

$$z_0 - 10z_1 - 10z_3 = 0$$

Subtracting the second equation from the first and the fourth from the third gives $25z_2 = 25$ and $40z_3 = 20$ so $z_2 = 1$ and $z_3 = 1/2$. Plugging in these values, we have two equations in two unknowns:

$$z_0 + 20z_1 = 20 \qquad z_0 - 10z_1 = 5$$

Taking differences, we conclude $30z_1 = 15$, so $z_1 = 1/2$, and $z_0 = 10$.

The reader may have already noticed that $z_0 = 10$ is the option price. This is no accident. What we have shown is that with $10 cash we can buy and sell shares of stock to produce the outcome of the option in all cases. In the terminology of Wall Street, $z_1 = 1/2, z_2 = 1, z_3 = 1/2$ is a **hedging strategy** that allows us to **replicate the option**. Once we can do this it follows that the fair price must be $10. To do this note that if we could sell it for $x > 10$ dollars, then we can take $10 of the cash to replicate the option and have a sure profit of $ $(x - 10)$.

6.2 Binomial Model

In this section we consider the Binomial model in which at each time the stock price the stock is multiplied by u (for "up") or multiplied by d (for "down"). As in the previous section we begin with the

6.2.1 One Period Case

There are two possible outcomes for the stock called heads (H) and tails (T).

$$S_0 \begin{array}{c} \nearrow S_1(H) = S_0 u \\ \\ \searrow S_1(T) = S_0 d \end{array}$$

We assume that there is an interest rate r, which means that \$1 at time 0 is the same as \$$(1+r)$ at time 1. For the model to be sensible, we need

$$0 < d < 1 + r < u. \tag{6.1}$$

Consider now an option that pays off $V_1(H)$ or $V_1(T)$ at time 1. This could be a call option $(S_1 - K)^+$, a put $(K - S_1)^+$, or something more exotic, so we will consider the general case. To find the "no arbitrage price" of this option we suppose we have V_0 in cash and Δ_0 shares of the stock at time 0, and want to pick these to match the option price exactly:

$$V_0 + \Delta_0 \left(\frac{1}{1+r} S_1(H) - S_0 \right) = \frac{1}{1+r} V_1(H) \tag{6.2}$$

$$V_0 + \Delta_0 \left(\frac{1}{1+r} S_1(T) - S_0 \right) = \frac{1}{1+r} V_1(T) \tag{6.3}$$

Notice that here we have to discount money at time 1 (i.e., divide it by $1 + r$) to make it comparable to dollars at time 0.

To find the values of V_0 and Δ_0 we define the risk neutral probability p^* so that

$$\frac{1}{1+r} (p^* S_0 u + (1 - p^*) S_0 d) = S_0 \tag{6.4}$$

Solving we have

$$p^* = \frac{1 + r - d}{u - d} \qquad 1 - p^* = \frac{u - (1 + r)}{u - d} \tag{6.5}$$

The conditions in (6.1) imply $0 < p^* < 1$.

Taking $p^*(6.2) + (1 - p^*)(6.3)$ and using (6.4) we have

$$V_0 = \frac{1}{1+r} (p^* V_1(H) + (1 - p^*) V_1(T)) \tag{6.6}$$

i.e., the value is the discounted expected value under the risk neutral probabilities. Taking the difference (6.2)–(6.3) we have

$$\Delta_0 \left(\frac{1}{1+r} (S_1(H) - S_1(T)) \right) = \frac{1}{1+r} (V_1(H) - V_1(T))$$

which implies that

$$\Delta_0 = \frac{V_1(H) - V_1(T)}{S_1(H) - S_1(T)} \tag{6.7}$$

To explain the notion of hedging we consider a concrete example.

Example 6.1. A stock is selling at \$60 today. A month from now it will either be at \$80 or \$50, i.e., $u = 4/3$ and $d = 5/6$. We assume an interest rate of $r = 1/18$ so the risk neutral probability is

$$\frac{19/18 - 5/6}{4/3 - 5/6} = \frac{4}{9}$$

Consider now a call option $(S_1 - 65)^+$. By (6.6) the value is

$$V_0 = \frac{18}{19} \cdot \frac{4}{9} \cdot 15 = \frac{120}{19} = 6.3158$$

Being a savvy businessman you offer to sell this for \$6.50. You are delighted when a customer purchases 10,000 calls for \$65,000, but then become worried about the fact that if the stock goes up you will lose \$85,000. By (6.7) the hedge ratio

$$\Delta_0 = \frac{15}{30} = 1/2$$

so you borrow \$300,000 − \$65,000 = \$235,000 and buy 5000 shares of stock.

Case 1. The stock goes up to \$80. Your stock is worth \$400,000. You have to pay \$150,000 for the calls and (19/18)\$235,000 = \$248,055 to redeem the loan so you make \$1,945 (in time 1 dollars).

Case 2. The stock drops to \$50. Your stock is worth \$250,000. You owe nothing for the calls but have to pay \$248,055 to redeem the loan so again you make \$1945.

The equality of the profits in the two cases may look like a miracle but it is not. By buying the correct amount of stock you replicated the option. This means you made a sure profit of the \$1842 difference (in time 0 dollars) between the selling price and fair price of the option, which translates into \$1945 time 1 dollars.

6.2.2 N Period Model

To solve the problem in general we work backwards from the end, repeatedly applying the solution of the one period problems. Let a be a string of H's and T's of length $n - 1$ which represents the outcome of the first $n - 1$ events. The value of the option at time n after the events in a have occurred, $V_n(a)$, and the amount of stock we need to hold in this situation, $\Delta_n(a)$, in order to replicate the option payoff satisfy:

$$V_n(a) + \Delta_n(a)\left(\frac{1}{1+r}S_{n+1}(aH) - S_n(a)\right) = \frac{1}{1+r}V_{n+1}(aH) \tag{6.8}$$

$$V_n(a) + \Delta_n(a)\left(\frac{1}{1+r}S_{n+1}(aT) - S_n(a)\right) = \frac{1}{1+r}V_{n+1}(aT) \tag{6.9}$$

Define the risk neutral probability $p_n^*(a)$ so that

$$S_n(a) = \frac{1}{1+r}[p_n^*(a)S_{n+1}(aH) + (1 - p_n^*(a))S_{n+1}(aT)] \qquad (6.10)$$

A little algebra shows that

$$p_n^*(a) = \frac{(1+r)S_n(a) - S_{n+1}(aT)}{S_{n+1}(aH) - S_{n+1}(aT)} \qquad (6.11)$$

In the binomial model one has $p_n^*(a) = (1+r-d)/(u-d)$. However, stock prices are not supposed to follow the binomial model, but are subject only to the no arbitrage restriction that $0 < p_n^*(a) < 1$. Notice that these probabilities depend on the time n and the history a.

Taking $p_n^*(a)(6.8) + (1 - p_n^*(a))(6.9)$ and using (6.10) we have

$$V_n(a) = \frac{1}{1+r}[p_n^*(a)V_{n+1}(aH) + (1 - p_n^*(a))V_{n+1}(aT)] \qquad (6.12)$$

i.e., the value is the discounted expected value under the risk neutral probabilities. Taking the difference (6.8)–(6.9) we have

$$\Delta_n(a)\left(\frac{1}{1+r}(S_{n+1}(aH) - S_{n+1}(aT))\right) = \frac{1}{1+r}(V_{n+1}(aH) - V_{n+1}(aT))$$

which implies that

$$\Delta_n(a) = \frac{V_{n+1}(aH) - V_{n+1}(aT)}{S_{n+1}(aH) - S_{n+1}(aT)} \qquad (6.13)$$

In words, $\Delta_n(a)$ is the ratio of the change in price of the option to the change in price of the stock. Thus for a call or put $|\Delta_n(a)| \leq 1$.

The option prices we have defined were motivated by the idea that by trading in the stock we could replicate the option exactly and hence they are the only price consistent with the absence of arbitrage. We will now go through the algebra needed to demonstrate this for the general n period model. Suppose we start with W_0 dollars and hold $\Delta_n(a)$ shares of stock between time n and $n+1$ when the outcome of the first n events is a. If we invest the money not in the stock in the money market account which pays interest r per period our wealth satisfies the recursion:

$$W_{n+1} = \Delta_n S_{n+1} + (1+r)(W_n - \Delta_n S_n) \qquad (6.14)$$

Here $\Delta_n S_{n+1}$ is the value of the stock we hold and $W_n - \Delta_n S_n$ is the amount of money we put in the bank at time n.

Theorem 6.3. *If $W_0 = V_0$ and we use the investment strategy in (6.13), then we have $W_n = V_n$.*

In words, we have a trading strategy that replicates the option payoffs.

Proof. We proceed by induction. By assumption the result is true when $n = 0$. Let a be a string of H and T of length n. (6.14) implies

$$W_{n+1}(aH) = \Delta_n(a)S_{n+1}(aH) + (1 + r)(W_n(a) - \Delta_n(a)S_n(a))$$
$$= (1 + r)W_n(a) + \Delta_n(a)[S_{n+1} - (1 + r)S_n(a)]$$

By induction the first term $= (1+r)V_n(a)$. Letting $q_n^*(a) = 1-p_n^*(a)$, (6.10) implies

$$(1 + r)S_n(a) = p_n^*(a)S_{n+1}(aH) + q_n^*(a)S_{n+1}(aT)$$

Subtracting this equation from $S_{n+1}(aH) = S_{n+1}(aH)$ we have

$$S_{n+1}(aH) - (1 + r)S_n(a) = q_n^*(a)[S_{n+1}(aH) - S_{n+1}(aT)]$$

Using (6.13) now, we have

$$\Delta_n(a)[S_{n+1} - (1 + r)S_n(a)] = q_n^*(a)[V_n(aH) - V_{n+1}(aT)]$$

Combining our results then using (6.12)

$$W_{n+1}(aH) = (1 + r)V_n(a) + q_n^*(a)[V_n(aH) - V_{n+1}(aT)]$$
$$= p_n^*(a)V_n(aH) + q_n^*V_n(aT) + q_n^*(a)[V_n(aH) - V_{n+1}(aT)]$$
$$= V_{n+1}(aH)$$

The proof that $W_{n+1}(aT) = V_{n+1}(aT)$ is almost identical.   □

Our next goal is to prove that the value of the option is its expected value under the risk neutral probability discounted by the interest rate (Theorem 6.5). The first step is

Theorem 6.4. *In the binomial model, under the risk neutral probability measure $M_n = S_n/(1 + r)^n$ is a martingale with respect to S_n.*

Proof. Let p^* and $1 - p^*$ be defined by (6.5). Given a string a of heads and tails of length n

$$P^*(a) = (p^*)^{H(a)}(1 - p^*)^{T(a)}$$

where $H(a)$ and $T(a)$ are the number of heads and tails in a. To check the martingale property we need to show that

$$E^*\left(\frac{S_{n+1}}{(1 + r)^{n+1}} \,\middle|\, S_n = s_n, \dots S_0 = s_0\right) = \frac{S_n}{(1 + r)^n}$$

where E^* indicates expected value with respect to P^*. Letting $X_{n+1} = S_{n+1}/S_n$ which is independent of S_n and is u with probability p^* and d with probability $1-p^*$ we have

$$E^* \left(\frac{S_{n+1}}{(1+r)^{n+1}} \,\middle|\, S_n = s_n, \ldots S_0 = s_0 \right)$$

$$= \frac{S_n}{(1+r)^n} E^* \left(\frac{X_{n+1}}{1+r} \,\middle|\, S_n = s_n, \ldots S_0 = s_0 \right) = \frac{S_n}{(1+r)^n}$$

since $E^* X_{n+1} = 1 + r$ by (6.10). $\square$

Notation. To make it easier to write computations like the last one we will let

$$E_n(Y) = E(Y|S_n = s_n, \ldots S_0 = s_0) \tag{6.15}$$

or in words, the conditional expectation of Y given the information at time n.

A second important martingale result is

Theorem 6.5. *Assume that the holdings $\Delta_n(a)$ are given by (6.13) and let W_n be the wealth process defined by (6.14). Under P^*, $W_n/(1+r)^n$ is a martingale, and hence the value has $V_0 = E^*(V_n/(1+r)^n)$.*

Proof. The second conclusion follows from the first and Theorem 6.3. A little arithmetic with (6.14) shows that

$$\frac{W_{n+1}}{(1+r)^{n+1}} = \frac{W_n}{(1+r)^n} + \Delta_n \left(\frac{S_{n+1}}{(1+r)^{n+1}} - \frac{S_n}{(1+r)^n} \right)$$

Since $\Delta_n(a)$ is an admissible gambling strategy and $S_n/(1+r)^n$ is a martingale, the desired result follows from Theorem 5.9. $\square$

6.3   Concrete Examples

Turning to examples, we will often use the following binomial model because it leads to easy arithmetic:

$$u = 2, \qquad d = 1/2, \qquad r = 1/4 \tag{6.16}$$

The risk neutral probabilities

$$p^* = \frac{1+r-d}{u-d} = \frac{5/4 - 1/2}{2 - 1/2} = \frac{1}{2}$$

and by (6.12) the option prices follow the recursion:

$$V_n(a) = .4[V_{n+1}(aH) + V_{n+1}(aT)] \tag{6.17}$$

Example 6.2 (Callback Options). In this option you can buy the stock at time 3 at its current price and then sell it at the highest price seen in the past for a profit of

$$V_3 = \max_{0 \le m \le 3} S_m - S_3$$

Our goal is to compute the value $V_n(a)$ and the replicating strategy $\Delta_n(a)$ for this option in the binomial model given in (6.16) with $S_0 = 4$. Here the numbers above the nodes are the stock price, while those below are the values of $V_n(a)$ and $\Delta_n(a)$. Starting at the right edge, $S_3(HTT) = 2$ but the maximum in the past is $8 = S_1(H)$ so $V_3(HTT) = 8 - 2 = 6$.

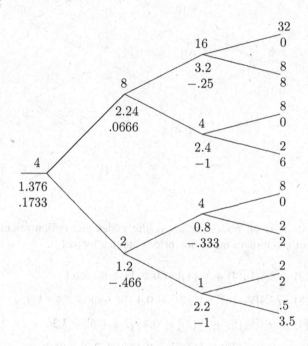

On the tree, stock prices are above the nodes and option prices below. To explain the computation of the option price note that by (6.17).

$$V_2(HH) = 0.4(V_3(HHH) + V_3(HHT)) = 0.4(0 + 8) = 3.2$$
$$V_2(HT) = 0.4(V_3(HTH) + V_3(HTT)) = 0.4(0 + 6) = 2.4$$
$$V_1(H) = 0.4(V_2(HH) + V_2(HT)) = 0.4(3.2 + 2.4) = 2.24$$

If one only wants the option price, then Theorem 6.5 which says that $V_0 = E^*(V_N/(1 + r)^N)$ is much quicker:

$$V_0 = (4/5)^3 \cdot \frac{1}{8} \cdot [0 + 8 + 0 + 6 + 0 + 2 + 2 + 2 + 3.5] = 1.376$$

Example 6.3 (Put Option). We will use the binomial model in (6.16) but now suppose $S_0 = 8$ and consider the put option with value $V_3 = (10 - S_3)^+$. The value of this option depends only on the price so we can reduce the tree considered above to:

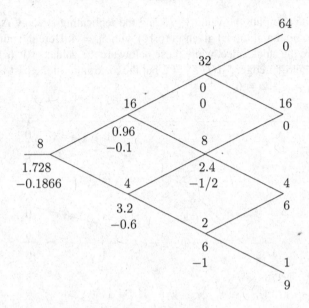

On the tree itself stock prices are above the nodes and option prices below. To explain the computation of the option price note that by (6.17).

$$V_2(2) = 0.4[V_3(4) + V_3(1)] = 0.4 \cdot [8 + 9] = 6.8$$
$$V_2(8) = 0.4[V_3(16) + V_3(2)] = 0.4 \cdot [0 + 6] = 2.4 \quad V_2(32) = 0$$
$$V_1(4) = 0.4[V_2(8) + V_2(2)] = 0.4 \cdot [2.4 + 6] = 3.36$$
$$V_1(16) = 0.4[V_2(32) + V_2(8)] = 0.4 \cdot [0 + 2.4] = 0.96$$
$$V_0(8) = 0.4[V_1(16) + V_1(40)] = 0.4 \cdot [0.96 + 3.36] = 1.728$$

Again if one only wants the option price, then Theorem 6.5 is much quicker:

$$V_0 = (4/5)^3 \cdot \left[6 \cdot \frac{3}{8} + 9 \cdot \frac{1}{8}\right] = 1.728$$

However if we want to compute the replicating strategy

$$\Delta_n(a) = \frac{V_{n+1}(aH) - V_{n+1}(aT)}{S_{n+1}(aH) - S_{n+1}(aT)}$$

one needs all of the information generated by the recursion.

$$\Delta_2(HH) = 0$$
$$\Delta_2(HT) = (0 - 6)/(16 - 4) = -0.5$$
$$\Delta_2(TT) = (6 - 9)/(4 - 1) = 1$$
$$\Delta_1(H) = (0 - 2.4)/(32 - 8) = -0.1$$
$$\Delta_1(T) = (2.4 - 6)/(8 - 2) = -0.6$$
$$\Delta_0(T) = (0.96 - 3.2)/(16 - 4)$$

Notice that $V_n(aH) \leq V_n(aT)$ and the change in the price of the option is always smaller than the change in the price of the stock so $-1 \leq \Delta_n(a) \leq 0$.

Example 6.4 (Put-Call Parity). Consider the binomial model with $S_0 = 32$, $u = 3/2$, $d = 2/3$, and $r = 1/6$. By (6.5) the risk neutral probability

$$p^* = \frac{1 + r - d}{u - d} = \frac{7/6 - 2/3}{3/2 - 2/3} = \frac{3/6}{5/6} = 0.6$$

so by (6.12) the value satisfies

$$V_n(a) = \frac{1}{7}(3.6 V_{n+1}(aH) + 2.4 V_{n+1}(aT))$$

We will now compute the values for the call and put with strike 49 and expiry 2. In the diagrams below the numbers above the line are the value of the stock and the ones below are the value of the option, and the replicating strategy.

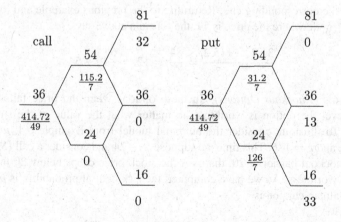

We have written the values as fractions to make it clear that the value at time 0 of these two options are exactly the same. Note that this is the only time they are equal.

Theorem 6.6. *The values V_P and V_C of the put and call options with the same strike K and expiration N are related by*

$$V_P - V_C = \frac{K}{(1+r)^N} - S_0$$

In particular if $K = (1+r)^N S_0$, then $V_P = V_C$.

Proof. The key observation is that

$$S_N + (K - S_N)^+ - (S_N - K)^+ = K$$

To prove this consider the two cases $S_N \geq K$ and $S_N \leq K$.

$$S_N + 0 - (S_N - K) = K$$
$$S_N + (K - S_N)- = K$$

Dividing our identity by $(1+r)^N$, taking E^* expected value and using the fact that $S_n/(1+r)^n$ is a martingale

$$S_0 + E^* \frac{(K - S_N)^+}{(1+r)^N} - E^* \frac{(K - S_N)^+}{(1+r)^N} = \frac{K}{(1+r)^N}$$

Since the second term on the left is V_P and the third is V_C the desired result follows. □

It is important to note that the last result can be used to compute the value of any put from the corresponding call. Returning to the previous example and looking at the time 1 node where the price is 54, the formula above says

$$\frac{31.2}{7} - \frac{115.2}{7} = \frac{84}{7} = 12 = \frac{6}{7}49 - 54$$

Example 6.5 (Knockout Options). In these options when the price falls below a certain level the option is worthless no matter what the value of the stock is at the end. To illustrate consider the binomial model from Example 6.4: $u = 3/2$, $d = 2/3$, and $r = 1/6$. This time we suppose $S_0 - 24$ and consider a call $(S_3 - 28)^+$ with a knockout barrier at 20, that is if the stock price drops below 20 the option becomes worthless. As we have computed the risk neutral probability is $p^* = 0.6$ and the value recursion is

$$V_n(a) = \frac{6}{7}[.6V_n(aH) + .4V_n(aT)],$$

with the extra boundary condition that if the price is ≤ 20 the value is 0.

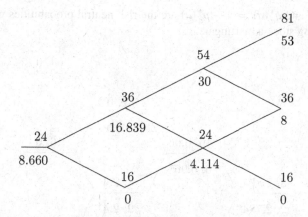

To check the answer note that the knockout feature eliminates one of the paths to 36 so

$$V_0 = (6/7)^3[(.6)^3 \cdot 53 + 2(.6)^2(.4) \cdot 8] = 8.660$$

From this we see that the knockout barrier reduced the value of the option by $(6/7)^3(0.6)^2(0.4) \cdot 8 = 0.7255$.

6.4 American Options

European option contracts specify an expiration date, and if the option is to be exercised at all, this must occur at the expiration date. An option whose owner can choose to exercise it at any time is called an **American option.** We will mostly be concerned with call and put options where the value at exercise is a function of the stock price, but it is no more difficult to consider path dependent options, so we derive the basic formulas in that generality.

Given a sequence a of heads and tails of length n let $g_n(a)$ be the value if we exercise at time n. Our first goal is to compute the value function $V_n(a)$ for the N-period problem. To simplify some statements, we will suppose (with essentially no loss of generality) that $g_N(a) \geq 0$, so $V_N(a) = g_N(a)$. If $g_n(a) < 0$, then we will not exercise the option so the value is $g_n(a)^+$.

To work backwards in time note that at time n we can exercise the option or let the game proceed for one more step. Since we will stop or continue depending on which choice gives the better payoff:

$$V_n(a) = \max\left\{g_n(a), \frac{1}{1+r}[p_n^*(a)V_{n+1}(aH) + q_n^*(a)V_{n+1}(aT)]\right\} \qquad (6.18)$$

where $p_n^*(a)$ and $q_n^*(a) = 1 - p_n^*(a)$ are the risk neutral probabilities which make the underlying stock a martingale.

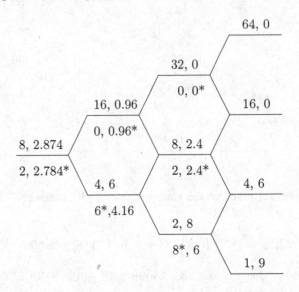

Example 6.6. For a concrete example, suppose as we did in Example 6.3 that the stock price follows the binomial model with $S_0 = 8$, $u = 2$, $d = 1/2$, $r = 1/4$ and consider a put option with strike 10, that is $g_n = (10 - s_n)^+$. The risk neutral probability $p^* = 0.5$ and the recursion is

$$V_{n-1}(a) = 0.4[V_n(aH) + V_n(aT)]$$

On the drawing above, the two numbers above each line are the price of the stock and the value of the option. Below the line are the value of the option if exercised, and the value computed by the recursion if we continue for one more period. A star indicates the large of the two, which is the value of the option at that time. To explain the solution, note that working backwards from the end.

$$V_2(2) = \max\{8, 0.4(6 + 9) = 6\} = 8$$
$$V_2(8) = \max\{2, 0.4(0 + 6) = 2.4\} = 2.4$$
$$V_2(32) = \max\{0, 0\} = 0$$
$$V_1(4) = \max\{6, 0.4(2.4 + 8) = 4.16\} = 6$$
$$V_1(16) = \max\{0, 0.4(0 + 2.4) = 0.96\} = 0.96$$
$$V_0(8) = \max\{2, 0.4(0.96 + 6) = 2.784\} = 2.784$$

This computes the value and the optimal strategy: stop or continue at each node depending on which value is larger. Notice that this is larger than the value 1.728

we computed for the European version. It certainly cannot be strictly less since one possibility in the American option is to continue each time and this turns it into a European option.

For some of the theoretical results it is useful to notice that

$$V_0 = \max_{\tau} E^* \left(\frac{g_\tau}{(1+r)^\tau} \right) \tag{6.19}$$

where the maximum is taken over all stopping times τ with $0 \leq \tau \leq N$. In Example 6.6 $\tau(T) = 1$ and $\tau(H) = 3$, i.e., if the stock goes down on the first step we stop. Otherwise we continue until the end.

$$V_0 = \frac{1}{2} \cdot 6 \cdot \frac{4}{5} + \frac{1}{8} \cdot 6 \cdot \left(\frac{4}{5} \right)^3 = 2.4 + 0.384$$

Proof. The key to prove the stronger statement

$$V_n(a) = \max_{\tau \geq n} E_n^* (g_\tau / (1+r)^{\tau-n})$$

where E_n^* is the conditional expectation given the events that have occurred up to time n. Let $W_n(a)$ denote the right-hand side. If we condition on the first n outcomes being a, then $P(\tau = n)$ is 1 or 0. In the first case we get $g_n(a)$. In the second case $W_n(a) = [p_n^*(a)W_{n+1}(aH) + q_n^*(a)W_{n+1}(aT)]/(1+r)$, so W_n and V_n satisfy the same recursion. □

Example 6.7. Continue now the set-up of the previous example but consider the call option $(S_n - 10)^+$. The computations are the same but the result is boring: the optimal strategy is to always continue, so there is no difference between the American and the European option.

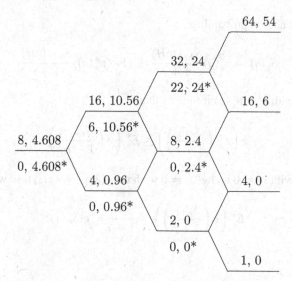

To spare the reader the chore of doing the arithmetic we give the recursion:

$$V_2(2) = \max\{0, 0\} = 0$$

$$V_2(8) = \max\{0, 0.4(0 + 6) = 2.4\} = 2.4$$

$$V_2(32) = \max\{22, 0.4(54 + 60 = 24\} = 24$$

$$V_1(4) = \max\{0, 0.4(0 + 2.4) = 0.96\} = 0.96$$

$$V_1(16) = \max\{6, 0.4(24 + 2.4) = 10.56\} = 10.56$$

$$V_0(8) = \max\{0, 0.4(10.56 + 0.96) = 4.608\} = 4.608$$

Our next goal is to prove that it is always optimal to continue in the case of the American call option. To explain the reason for this we formulate an abstract result. We say that g is convex if whenever $0\lambda \leq 1$ and s_1, s_2 are real numbers

$$g(\lambda s_1 + (1 - \lambda)s_2) \leq \lambda g(s_1) + (1 - \lambda)g(s_2) \qquad (6.20)$$

Geometrically the line segment from $(s_1, g_1(s))$ to $(s_2, g_2(s))$ always lies above the graph of the function g. This is true for the call $g(x) = (x - K)^+$ and the put $g(x) = (K - x)^+$. However only the call satisfies all of the conditions in the following result:

Theorem 6.7. *If g is a nonnegative convex function with $g(0) = 0$, then for the American option with payoff $g(S_n)$ it is optimal to wait until the end of exercise.*

Proof. Since $S_n/(1 + r)^n$ is a martingale under P^*

$$g(S_n) = g\left(E_n^*\left(\frac{S_{n+1}}{1 + r}\right)\right)$$

Under the risk neutral probability

$$S_n(a) = p_n^*(a)\frac{S_{n+1}(aH)}{1 + r} + (1 - p_n^*(a))\frac{S_{n+1}(aH)}{1 + r}$$

Using (6.20) with $\lambda = p_n^*(a)$ it follows that

$$g\left(E_n^*\left(\frac{S_{n+1}}{1 + r}\right)\right) \leq E_n^*\left(g\left(\frac{S_{n+1}}{1 + r}\right)\right)$$

Using (6.20) with $s_2 = 0$ and $g(0) = 0$ we have $g(\lambda s_1) \leq \lambda g(s_1)$, so we have

$$E_n^*\left(g\left(\frac{S_{n+1}}{1 + r}\right)\right) \leq \frac{1}{1 + r}E_n^* g(S_{n+1})$$

Combining three of the last four equations we have

$$g(S_n) \leq \frac{1}{1+r} E_n^* g(S_{n+1})$$

This shows that if we were to stop at time n for some outcome a, it would be better to continue. Using (6.19) now the desired result follows. $\qquad\square$

6.5 Black–Scholes Formula

Many options take place over a time period of one or more months, so it is natural consider S_t to be the stock price after t years. We could use a binomial model in which prices change at the end of each day but it would also be natural to update prices several times during the day. Let h be the amount of time measured in years between updates of the stock price. This h will be very small, e.g., $1/365$ for daily updates so it is natural to let $h \to 0$. Knowing what will happen when we take the limit we will let

$$S_{nh} = S_{(n-1)h} \exp(\mu h + \sigma \sqrt{h} X_n)$$

where $P(X_n = 1) = P(X_n = -1) = 1/2$. This is binomial model with

$$u = \exp(\mu h + \sigma \sqrt{h}) \qquad d = \exp(\mu h - \sigma \sqrt{h}) \qquad (6.21)$$

Iterating we see that

$$S_{nh} = S_0 \exp\left(\mu n h + \sigma \sqrt{h} \sum_{m=1}^{n} X_m\right) \qquad (6.22)$$

If we let $t = nh$ the first term is just μt. Writing $h = t/n$ the second term becomes

$$\sigma \sqrt{t} \cdot \frac{1}{\sqrt{n}} \sum_{m=1}^{n} X_m$$

To take the limit as $n \to \infty$, we use the

Theorem 6.8. Central Limit Theorem. Let $X_1, X_2, \ldots$ be i.i.d. with $EX_i = 0$ and $\mathrm{var}(X_i) = 1$ Then for all x we have

$$P\left(\frac{1}{\sqrt{n}} \sum_{m=1}^{n} X_m \leq x\right) \to P(\chi \leq x) \qquad (6.23)$$

where χ has a standard normal distribution. That is,

$$P(\chi \le x) = \int_{-\infty}^{x} \frac{1}{\sqrt{2\pi}} e^{-y^2/2} \, dy$$

The conclusion in (6.23) is often written as

$$\frac{1}{\sqrt{n}} \sum_{m=1}^{n} X_m \Rightarrow \chi$$

where $\Rightarrow$ is read "converges in distribution to." Recalling that if we multiply a standard normal χ by a constant c then the result has a normal distribution with mean 0 and variance σ^2, we see that

$$\sqrt{t} \cdot \frac{1}{\sqrt{n}} \sum_{m=1}^{n} X_m \Rightarrow \sqrt{t}\chi$$

and the limit is a normal with mean 0 and variance t.

This motivates the following definition:

Definition. $B(t)$ is a standard Brownian motion if $B(0) = 0$ and it satisfies the following conditions:

(a) *Independent increments.* Whenever $0 = t_0 < t_1 < \ldots < t_k$

$$B(t_1) - B(t_0), \ldots, B(t_k) - B(t_{k-1}) \quad \text{are independent.}$$

(b) *Stationary increments.* The distribution of $B_t - B_s$ is normal $(0, t - s)$.
(c) $t \to B_t$ is continuous.

To explain (a) note that if $n_i = t_i/h$, then the sums

$$\sum_{n_{i-1} < m \le n_i} X_m \quad i = 1, \ldots k$$

are independent. For (b) we note that the distribution of the sum only depends on the number of terms and use the previous calculation. Condition (c) is a natural assumption for the physical system which motivated this definition: the erratic movement of a pollen grain in water as seen under a microscope by Brown in 1825.

Using the new definition, our stock price model can be written as

$$S_t = S_0 \cdot \exp(\mu t + \sigma B_t) \tag{6.24}$$

where B_t is a standard Brownian motion Here μ is the **exponential growth** rate of the stock and σ is its **volatility**. If we also assume that the per period interest rate in the approximating model is rh, and recall that

$$\left(\frac{1}{1 + rh} \right)^{t/h} = \frac{1}{(1 + rh)^{t/h}} \to \frac{1}{e^{rt}} = e^{-rt}$$

then the discounted stock price is

$$e^{-rt}S_t = S_0 \cdot \exp((\mu - r)t + \sigma B_t)$$

By the formula for the moment generating function for the normal with mean 0 and variance $\sigma^2 t$, see (5.15),

$$E \exp(-(\sigma^2/2)t + \sigma B_t) = 1$$

Since B_t has independent increments, if we let

$$\mu = r - \sigma^2/2 \qquad (6.25)$$

then reasoning as for the exponential martingale, Example 5.5, the discounted stock price, $e^{-rt}S_t$ is a martingale.

Extrapolating wildly from discrete time, we can guess that the option price is its expected value after changing the probabilities to make the stock price a martingale.

Theorem 6.9. *Write E^* for expected values when $\mu = r - \sigma^2/2$ in (6.24). The value of a European option $g(S_T)$ is given by $E^* e^{-rT} g(S_T)$.*

Proof. We prove this by taking limits of the discrete approximation. The risk neutral probabilities, p_h^*, are given by

$$p_h^* = \frac{1 + rh - d}{u - d}. \qquad (6.26)$$

Using the formulas for u and d in (6.21 and recalling that $e^x = 1 + x + x^2/2 + \cdots$,

$$u = 1 + \mu h + \sigma \sqrt{h} + \frac{1}{2}(\mu h + \sigma \sqrt{h})^2 + \ldots$$

$$= 1 + \sigma \sqrt{h} + (\sigma^2/2 + \mu)h + \ldots \qquad (6.27)$$

$$d = 1 + \mu h - \sigma \sqrt{h} + (\sigma^2/2 + \mu)h + \ldots$$

so from (6.26) we have

$$p_h^* \approx \frac{\sigma \sqrt{h} + (r - \mu - \sigma^2/2)h}{2\sigma \sqrt{h}} = \frac{1}{2} + \frac{r - \mu - \sigma^2/2}{2\sigma} \sqrt{h}$$

If $X_1^h, X_2^h, \ldots$ are i.i.d. with

$$P(X_1^h = 1) = p_h^* \qquad P(X_1^h = -1) = 1 - p_h^*$$

then the mean and variance are

$$EX_i^h = 2p_h^* = \frac{(r - \mu - \sigma^2/2)}{\sigma} \sqrt{h}$$

$$\text{var}\,(X_i^h) = 1 - (EX_i^h)^2 \to 1$$

To apply the central limit theorem we note that

$$\sigma \sqrt{h} \sum_{m=1}^{t/h} X_m^h = \sigma \sqrt{h} \sum_{m=1}^{t/h} (X_m^h - EX_m^h) + \sigma \sqrt{h} \sum_{m=1}^{t/h} EX_m^h$$

$$\to \sigma B_t + (r - \mu + \sigma^2/2)t$$

so under the risk neutral measure, P^*,

$$S_t = S_0 \cdot \exp((r - \sigma^2/2)t + \sigma B_t)$$

The value of the option $g(S_T)$ in the discrete approximation is given by the expected value under its risk neutral measure. Ignoring the detail of proving that the limit of expected values is the expected value of the limit, we have proved the desired result.

$\square$

6.5.1 The Black–Scholes Partial Differential Equation

We continue to suppose that the option payoff at time T is $g(S_T)$. Let $V(t, s)$ be the value of the option at time $t < T$ when the stock price is s. Reasoning with the discrete time approximation and ignoring the fact that the value in this case depends on h,

$$V(t - h, s) = \frac{1}{1 + rh} \left[p^* V(t, su) + (1 - p^*) V(t, sd) \right]$$

Doing some algebra we have

$$V(t, s) - (1 + rh)V(t - s, h) = p^* [V(t, s) - V(t, su)]$$
$$+ (1 - p^*)[V(t, s) - V(t, sd)]$$

Dividing by h we have

$$\frac{V(t, s) - V(t - h, s)}{h} - rV(t - h, s) \tag{6.28}$$

$$= p^* \left[\frac{V(t, s) - V(t, su)}{h} \right] + (1 - p^*) \left[\frac{V(t, s) - V(t, sd)}{h} \right]$$

Letting $h \to 0$ the left-hand side of (6.28) converges to

$$\frac{\partial V}{\partial t}(t,s) - rV(t,s) \qquad (6.29)$$

Expanding $V(t,s)$ in a power series in s

$$V(t,s') - V(t,s) \approx \frac{\partial V}{\partial x}(t,s)(s'-s) + \frac{\partial^2 V}{\partial x^2}(t,s)\frac{(s'-s)^2}{2}$$

Using the last equations with $s' = su$ and $s' = sd$, the right-hand side of (6.28) is

$$\approx \frac{\partial V}{\partial x}(t,s)s[(1-u)p^* + (1-d)(1-p^*)]/h$$

$$-\frac{1}{2}\frac{\partial^2 V}{\partial x^2}(t,s)s^2[p^*(1-u)^2 + (1-p^*)(1-d)^2]/h$$

From (6.26)

$$\frac{(1-u)p^* + (1-d)(1-p^*)}{h} \approx -\left(\frac{\sigma^2}{2} + \mu\right) = -r$$

$$\frac{(1-u)^2p^* + (1-d)^2(1-p^*)}{h} \approx \sigma^2$$

so taking the limit, the right-hand side of (6.28) is

$$\frac{\partial V}{\partial x}(t,s)s[-rh] + \frac{1}{2}\frac{\partial^2 V}{\partial x^2}(t,s)s^2\sigma^2 h$$

Combining the last equation with (6.29) and (6.28) we have that the value function satisfies

$$\frac{\partial V}{\partial t} - rV(t,s) + rs\frac{\partial V}{\partial x}(t,s) + \frac{1}{2}\sigma^2 s^2 \frac{\partial^2 V}{\partial x^2}(t,s) = 0 \qquad (6.30)$$

for $0 \le t < T$ with boundary condition $V(T,s) = g(s)$.

6.6 Calls and Puts

We will now apply the theory developed in the previous section to the concrete examples of calls and puts. At first glance the formula in the first result may look complicated, but given that the value is defined by solving a PDE, it is remarkable that such a simple formula exists.

Theorem 6.10. *The price of the European call option* $(S_t - K)^+$ *is given by*

$$S_0\Phi(d_1) - e^{-rt}K\Phi(d_2)$$

where the constants

$$d_1 = \frac{\ln(S_0/K) + (r + \sigma^2/2)t}{\sigma\sqrt{t}} \qquad d_2 = d_1 - \sigma\sqrt{t}.$$

Proof. Using the fact that $\log(S_t/S_0)$ has a normal$(\mu t, \sigma^2 t)$ distribution with $\mu = r - \sigma^2/2$, we see that

$$E^*(e^{-rt}(S_t - K)^+) = e^{-rt}\int_{\log(K/S_0)}^{\infty}(S_0 e^y - K)\frac{1}{\sqrt{2\pi\sigma^2 t}}e^{-(y-\mu t)^2/2\sigma^2 t}\,dy$$

Splitting the integral into two and then changing variables $y = \mu t + w\sigma\sqrt{t}$, $dy = \sigma\sqrt{t}\,dw$ the integral is equal to

$$= e^{-rt}S_0 e^{\mu t}\frac{1}{\sqrt{2\pi}}\int_{\alpha}^{\infty}e^{w\sigma\sqrt{t}}e^{-w^2/2}\,dw - e^{-rt}K\frac{1}{\sqrt{2\pi}}\int_{\alpha}^{\infty}e^{-w^2/2}\,dw \qquad (6.31)$$

where $\alpha = (\log(K/S_0) - \mu t)/\sigma\sqrt{t}$. To handle the first term, we note that

$$\frac{1}{\sqrt{2\pi}}\int_{\alpha}^{\infty}e^{w\sigma\sqrt{t}}e^{-w^2/2}\,dw = e^{t\sigma^2/2}\int_{\alpha}^{\infty}\frac{1}{\sqrt{2\pi}}e^{-(w-\sigma\sqrt{t})^2/2}\,dw$$

$$= e^{t\sigma^2/2}\,P(\text{normal}(\sigma\sqrt{t}, 1) > \alpha)$$

The last probability can be written in terms of the distribution function Φ of a normal(0,1) χ, i.e., $\Phi(t) = P(\chi \le t)$, by noting

$$P(\text{normal}(\sigma\sqrt{t}, 1) > \alpha) = P(\chi > \alpha - \sigma\sqrt{t})$$

$$= P(\chi \le \sigma\sqrt{t} - \alpha) = \Phi(\sigma\sqrt{t} - \alpha)$$

where in the middle equality we have used the fact that χ and $-\chi$ have the same distribution. Using the last two computations in (6.31) converts it to

$$e^{-rt}S_0 e^{\mu t}e^{\sigma^2 t/2}\Phi(\sigma\sqrt{t} - \alpha) - e^{-rt}K\Phi(-\alpha)$$

Now $e^{-rt}e^{\mu t}e^{\sigma^2 t/2} = 1$ since $\mu = r - \sigma^2/2$. As for the argument of the first normal

$$d_1 = \sigma\sqrt{t} - \alpha = \frac{\log(S_0/K) + (r - \sigma^2/2)t}{\sigma\sqrt{t}} + \sigma\sqrt{t}$$

which agrees with the formula given in the theorem. The second one is much easier to see: $d_2 = d_1 - \sigma\sqrt{t}$. □

Example 6.8 (A Google Call Options). On the morning of December 5, 2011, Google stock was selling for $620 a share and a March 12 call option with strike $K = 635$ was selling for $33.10. To compare this with the prediction of the Black–Scholes formula we assume an interest rate of $r = 0.01$ per year and assume a volatility $\sigma = 0.3$. The 100 days until expiration of the option are $t = 0.27393$ years. With the help of a little spreadsheet we find that the formula predicts a price of $32.93.

Example 6.9. **Put-call parity** allows us to compute the value of the put option, V_P from the value of the call option V_C by the formula:

$$V_P - V_C = e^{-rT} K - S_0$$

In the example for March 12 Google options, $\exp(-rt) = 0.9966$ so we might as well ignore that factor. As the next table shows the formula works well in practice

strike	V_P	V_C	$S_0 + V_P - V_C$
600	28.00	55.60	598.20
640	45.90	32.70	638.20
680	74.16	17.85	681.31

6.7 Exercises

6.1. A stock is now at $110. In a year its price will either be $121 or $99. (a) Assuming that the interest rate is $r = 0.04$ find the price of a call $(S_1 - 113)+$. (b) How much stock Δ_0 do we need to buy to replicate the option. (c) Verify that having V_0 in cash and Δ_0 in stock replicates the option exactly.

6.2. A stock is now at $60. In a year its price will either be $75 or $45. (a) Assuming that the interest rate is $r = 0.05$ find the price of a put $(60 - S_1-)+$. (b) How much stock Δ_0 do we need to sell to replicate the option. (c) Verify that having V_0 in cash and Δ_0 in stock replicates the option exactly.

6.3. It was crucial for our no arbitrage computations that there were only two possible values of the stock. Suppose that a stock is now at 100, but in one month may be at 130, 110, or 80 in outcomes that we call 1, 2, and 3. (a) Find all the (nonnegative) probabilities p_1, p_2, and $p_3 = 1 - p_1 - p_2$ that make the stock price a martingale. (b) Find the maximum and minimum values, v_1 and v_0, of the expected value of the call option $(S_1 - 105)^+$ among the martingale probabilities. (c) Show that we can start with v_1 in cash, buy x_1 shares of stock and we have $v_1 + x_1(S_1 - S_0) \geq (S_1 - 105)^+$ in all three outcomes with equality for 1 and 3. (d) If we start with v_0 in cash, buy x_0 shares of stock and we have $v_0 + x_0(S_1 - S_0) \leq (S_1 - 105)^+$ in all three outcomes with equality for 2 and 3. (e) Use (c) and (d) to argue that the only prices for the option consistent with absence of arbitrage are those in $[v_0, v_1]$.

6.4. The Cornell hockey team is playing a game against Harvard that it will either win, lose, or draw. A gambler offers you the following three payoffs, each for a $1 bet:

	win	lose	draw
Bet 1	0	1	1.5
Bet 2	2	2	0
Bet 3	.5	1.5	0

(a) Assume you are able to buy any amounts (even negative) of these bets. Is there an arbitrage opportunity? (b) What if only the first two bets are available?

6.5. Suppose Microsoft stock sells for 100 while Netscape sells for 50. Three possible outcomes of a court case will have the following impact on the two stocks:

	Microsoft	Netscape
1 (win)	120	30
2 (draw)	110	55
3 (lose)	84	60

What should we be willing to pay for an option to buy Netscape for 50 after the court case is over? Answer this question in two ways: (i) find a probability distribution so that the two stocks are martingales, (ii) show that by using cash and buying Microsoft and Netscape stock one can replicate the option.

6.6. Consider the two-period binomial model with $u = 2, d = 1/2$ and interest rate $r = 1/4$ and suppose $S_0 = 100$. What is the value of the European call option with strike price 80, i.e., the option with payoff $(S_2 - 80)^+$. Find the stock holdings Δ_0, $\Delta_1(H)$, and $\Delta_1(T)$ need to replicate the option exactly.

6.7. Consider the two-period binomial model with $u = 3/2, d = 2/3$, interest rate $r = 1/6$ and suppose $S_0 = 45$. What is the value of the European call option with strike price 50, i.e., the option with payoff $(50 - S_2)^+$. Find the stock holdings Δ_0, $\Delta_1(H)$, and $\Delta_1(T)$ need to replicate the option exactly.

6.8. The payoff of the Asian option is based on the average price: $A_n = (S_0 + \cdots + S_n)/(n+1)$. Suppose that the stock follows the binomial model with $S_0 = 4, u = 2$, $d = 1/2$, and $r = 1/4$. (a) Compute the value function $V_n(a)$ and the replicating portfolio $\Delta_n(a)$ for the three period call option with strike 4. (b) Check your answer for V_0 by using $V_0 = E^*(V_3/(1+r)^3)$.

6.9. In the putback option at time 3 you can buy the stock for the lowest price seen in the past and sell it at its current price for a profit of

$$V_3 = S_3 - \min_{0 \le m \le 3} S_m$$

Suppose that the stock follows the binomial model with $S_0 = 4$, $u = 2$, $d = 1/2$, and $r = 1/4$. (a) Compute the value function $V_n(a)$ and the replicating portfolio $\Delta_n(a)$ for the three period call option with strike 4. (b) Check your answer for V_0 by using $V_0 = E^*(V_3/(1+r)^3)$.

6.10. Consider the three-period binomial model with $u = 3$, $d = 1/2$ and $r = 1/3$ and $S_0 = 16$. The European prime factor option pays off \$1 for each factor in the prime factorization of the stock price at time 3 (when the option expires). For example, if the stock price is $24 = 2^3 3^1$, then the payoff is $4 = 3 + 1$. Find the no arbitrage price of this option.

6.11. Suppose $S_0 = 27$, $u = 4/3$, $d = 2/3$ and interest rate $r = 1/9$. The European "cash-or-nothing option" pays \$1 if $S_3 > 27$ and 0 otherwise. Find the value of the option V_n and for the hedge Δ_n.

6.12. Assume the binomial model with $S_0 = 54$, $u = 3/2$, $d = 2/3$, and $r = 1/6$. and consider a put $(50 - S_3)^+$ with a knockout barrier at 70. Find the value of the option.

6.13. Consider now a four period binomial model with $S_0 = 32$, $u = 2$, $d = 1/2$, and $r = 1/4$, and suppose we have a put $(50 - S_4)^+$ with a knockout barrier at 100. Show that the knockout option as the same value as an option that pays off $(50 - S_4)^+$ when $S_4 = 2$, 8, or 32, 0 when $S_4 = 128$, and -18 when $S_4 = 512$. (b) Compute the value of the option in (a).

6.14. Consider the binomial model with $S_0 = 64$, $u = 2$, $d = 1/2$, and $r = 1/4$. (a) Find the value $V_n(a)$ of the call option $(S_3 - 125)^+$ and the hedging strategy $\Delta_n(a)$. (b) Check your answer to (a) by computing $V_0 = E^*(V_3/(1+r)^3)$. (c) Find the value at time 0 of the put option.

6.15. Consider the binomial model with $S_0 = 27$, $u = 4/3$, $d = 2/3$, and $r = 1/9$. (a) Find the risk neutral probability p^*. (b) Find value $V_n(a)$ of the put option $(30 - S_3)^+$ and the hedging strategy $\Delta_n(a)$. (c) Check your answer to (b) by computing $V_0 = E^*(V_3/(1+r)^3)$.

6.16. Consider the binomial model of Problem 6.15 $S_0 = 27$, $u = 4/3$, $d = 2/3$, and $r = 1/9$ but now (a) find value and the optimal exercise strategy for the American put option $(30 - S_3)^+$, and (b) find the value of the American call option $(S_3 - 30)^+$.

6.17. Continuing with the model of previous problem $S_0 = 27$, $u = 4/3$, $d = 2/3$, and $r = 1/9$, we are now interested in finding value V_S of the American straddle $|S_3 - 30|$. Comparing with the values V_P and V_C of the call and the put computed in the previous problem we see that $V_S \leq V_P + V_C$. Explain why this should be true.

6.18. Consider the three-period binomial model with $S_0 = 16$, $u = 3$, $d = 1/2$ and $r = 1/3$. An American limited liability call option pays $\min\{(S_n - 10)^+, 60\}$ if exercised at time $0 \leq n \leq 3$. In words it is a call option but your profit is limited to \$60. Find the value and the optimal exercise strategy.

6.19. In the American version of the callback option, you can buy the stock at time n at its current price and then sell it at the highest price seen in the past for a profit of $V_n = \max_{0 \le m \le n} S_m - S_n$. Compute the value of the three period version of this option when the stock follows the binomial model with $S_0 = 8$, $u = 2$, $d = 1/2$, and $r = 1/4$.

6.20. The payoff of the Asian option is based on the average price: $A_n = (S_0 + \cdots + S_n)/(n + 1)$. Suppose that the stock follows the binomial model with $S_0 = 4$, $u = 2$, $d = 1/2$, and $r = 1/4$. Find the value of the American version of the three period Asian option, $(S_n - 4)^+$, i.e., when you can exercise the option at any time.

6.21. Show that for any a and b, $V(s, t) = as + be^{rt}$ satisfies the Black–Scholes differential equation. What investment does this correspond to?

6.22. Find a formula for the value (at time 0) of cash-or-nothing option that pays off \$1 if $S_t > K$ and 0 otherwise. What is the value when the strike is the initial value, the option is for $1/4$ year, the volatility is $\sigma = 0.3$, and for simplicity we suppose that the interest rate is 0.

6.23. On May 22, 1998, Intel was selling at 74.625. Use the Black–Scholes formula to compute the value of a January 2000 call ($t = 1.646$ years) with strike 100, assuming the interest rate was $r = 0.05$ and the volatility $\sigma = 0.375$.

6.24. On December 20, 2011, stock in Kraft Foods was selling at 36.83. (a) Use the Black–Scholes formula to compute the value of a March 12 call ($t = 0.227$ years) with strike 33, assuming an interest rate of $r = 0.01$ and the volatility $\sigma = 0.15$. The volatility here has been chosen to make the price consistent with the bid-ask spread of (3.9,4.0). (b) Is the price of 0.4 for a put with strike 33 consistent with put-call parity?

6.25. On December 20, 2011, stock in Exxon Mobil was selling at 81.63. (a) Use the Black–Scholes formula to compute the value of an April 12 call ($t = 0.3123$ years) with strike 70, assuming an interest rate of $r = 0.01$ and the volatility $\sigma = 0.26$. The volatility here has been chosen to make the price consistent with the bid-ask spread of (12.6,12.7). (b) Is the price of 1.43 for a put with strike 70 consistent with put-call parity?

Appendix A
Review of Probability

Here we will review some of the basic facts usually taught in a first course in probability, concentrating on the ones that are important in the book.

A.1 Probabilities, Independence

The term **experiment** is used to refer to any process whose outcome is not known in advance. Two simple experiments are flip a coin, and roll a die. The **sample space** associated with an experiment is the set of all possible outcomes. The sample space is usually denoted by Ω, the capital Greek letter Omega.

Example A.1 (Flip Three Coins). The flip of one coin has two possible outcomes, called "Heads" and "Tails," and denoted by H and T. Flipping three coins leads to $2^3 = 8$ outcomes:

$$
\begin{array}{cccc}
 & HHT & HTT & \\
HHH & HTH & THT & TTT \\
 & THH & TTH &
\end{array}
$$

Example A.2 (Roll Two Dice). The roll of one die has six possible outcomes: 1, 2, 3, 4, 5, 6. Rolling two dice leads to $6^2 = 36$ outcomes $\{(m, n) : 1 \le m, n \le 6\}$.

The goal of probability theory is to compute the probability of various events of interest. Intuitively, an event is a statement about the outcome of an experiment. Formally, an **event** is a subset of the sample space. An example for flipping three coins is "two coins show Heads," or

$$
A = \{HHT, HTH, THH\}
$$

© Springer International Publishing Switzerland 2016
R. Durrett, *Essentials of Stochastic Processes*, Springer Texts in Statistics,
DOI 10.1007/978-3-319-45614-0

An example for rolling two dice is "the sum is 9," or

$$B = \{(6,3), (5,4), (4,5), (3,6)\}$$

Events are just sets, so we can perform the usual operations of set theory on them. For example, if $\Omega = \{1,2,3,4,5,6\}$, $A = \{1,2,3\}$, and $B = \{2,3,4,5\}$, then the **union** $A \cup B = \{1,2,3,4,5\}$, the **intersection** $A \cap B = \{2,3\}$, and the **complement of** A, $A^c = \{4,5,6\}$. To introduce our next definition, we need one more notion: two events are **disjoint** if their intersection is the empty set, $\emptyset$. A and B are not disjoint, but if $C = \{5,6\}$, then A and C are disjoint.

A **probability** is a way of assigning numbers to events that satisfies:

(i) For any event A, $0 \le P(A) \le 1$.
(ii) If Ω is the sample space, then $P(\Omega) = 1$.
(iii) For a finite or infinite sequence of disjoint events $P(\cup_i A_i) = \sum_i P(A_i)$.

In words, the probability of a union of disjoint events is the sum of the probabilities of the sets. We leave the index set unspecified since it might be finite,

$$P(\cup_{i=1}^{k} A_i) = \sum_{i=1}^{k} P(A_i)$$

or it might be infinite, $P(\cup_{i=1}^{\infty} A_i) = \sum_{i=1}^{\infty} P(A_i)$.

In Examples A.1 and A.2, all outcomes have the same probability, so

$$P(A) = |A|/|\Omega|$$

where $|B|$ is short for the number of points in B. For a very general example of a probability, let $\Omega = \{1,2,\ldots,n\}$; let $p_i \ge 0$ with $\sum_i p_i = 1$; and define $P(A) = \sum_{i \in A} p_i$. Two basic properties that follow immediately from the definition of a probability are

$$P(A) = 1 - P(A^c) \tag{A.1}$$

$$P(B \cup C) = P(B) + P(C) - P(B \cap C) \tag{A.2}$$

To illustrate their use consider the following:

Example A.3. Roll two dice and suppose for simplicity that they are red and green. Let $A = $ "at least one 4 appears," $B = $ "a 4 appears on the red die," and $C = $ "a 4 appears on the green die," so $A = B \cup C$.

Solution 1. $A^c = $ "neither die shows a 4," which contains $5 \cdot 5 = 25$ outcomes so (A.1) implies $P(A) = 1 - 25/36 = 11/36$.

Solution 2. $P(B) = P(C) = 1/6$ while $P(B \cap C) = P(\{4,4\}) = 1/36$, so (A.2) implies $P(A) = 1/6 + 1/6 - 1/36 = 11/36$.

A.1.1   Conditional Probability

Suppose we are told that the event A with $P(A) > 0$ occurs. Then the sample space is reduced from Ω to A and the probability that B will occur given that A has occurred is

$$P(B|A) = P(B \cap A)/P(A) \tag{A.3}$$

To explain this formula, note that (i) only the part of B that lies in A can possibly occur, and (ii) since the sample space is now A, we have to divide by $P(A)$ to make $P(A|A) = 1$. Multiplying on each side of (A.3) by $P(A)$ gives us the **multiplication rule**:

$$P(A \cap B) = P(A)P(B|A) \tag{A.4}$$

Intuitively, we think of things occurring in two stages. First we see if A occurs, then we see what the probability B occurs given that A did. In many cases these two stages are visible in the problem.

Example A.4. Suppose we draw two balls without replacement from an urn with six blue balls and four red balls. What is the probability we will get two blue balls? Let A = blue on the first draw, and B = blue on the second draw. Clearly, $P(A) = 6/10$. After A occurs, the urn has five blue balls and four red balls, so $P(B|A) = 5/9$ and it follows from (A.4) that

$$P(A \cap B) = P(A)P(B|A) = \frac{6}{10} \cdot \frac{5}{9}$$

To see that this is the right answer notice that if we draw two balls without replacement and keep track of the order of the draws, then there are $10 \cdot 9$ outcomes, while $6 \cdot 5$ of these result in two blue balls being drawn.

The multiplication rule is useful in solving a variety of problems. To illustrate its use we consider:

Example A.5. Suppose we roll a four-sided die then flip that number of coins. What is the probability we will get exactly one Heads? Let B = we get exactly one Heads, and A_i = an i appears on the first roll. Clearly, $P(A_i) = 1/4$ for $1 \leq i \leq 4$. A little thought gives

$$P(B|A_1) = 1/2, \quad P(B|A_2) = 2/4, \quad P(B|A_3) = 3/8, \quad P(B|A_4) = 4/16$$

so breaking things down according to which A_i occurs,

$$P(B) = \sum_{i=1}^{4} P(B \cap A_i) = \sum_{i=1}^{4} P(A_i)P(B|A_i)$$

$$= \frac{1}{4}\left(\frac{1}{2} + \frac{2}{4} + \frac{3}{8} + \frac{4}{16}\right) = \frac{13}{32}$$

One can also ask the reverse question: if B occurs, what is the most likely cause? By the definition of conditional probability and the multiplication rule,

$$P(A_i|B) = \frac{P(A_i \cap B)}{\sum_{j=1}^{4} P(A_j \cap B)} = \frac{P(A_i)P(B|A_i)}{\sum_{j=1}^{4} P(A_j)P(B|A_j)} \qquad (A.5)$$

This little monster is called **Bayes' formula**, but it will not see much action here.

Last but far from least, two events A and B are said to be **independent** if $P(B|A) = P(B)$. In words, knowing that A occurs does not change the probability that B occurs. Using the multiplication rule this definition can be written in a more symmetric way as

$$P(A \cap B) = P(A) \cdot P(B) \qquad (A.6)$$

Example A.6. Roll two dice and let $A =$ "the first die is 4."

Let $B_1 =$ "the second die is 2." This satisfies our intuitive notion of independence since the outcome of the first dice roll has nothing to do with that of the second. To check independence from (A.6), we note that $P(B_1) = 1/6$ while the intersection $A \cap B_1 = \{(4,2)\}$ has probability 1/36.

$$P(A \cap B_1) = \frac{1}{36} \neq \frac{1}{6} \cdot \frac{4}{36} = P(A)P(B_1)$$

Let $B_2 =$ "the sum of the two dice is 3." The events A and B_2 are disjoint, so they cannot be independent:

$$P(A \cap B_2) = 0 < P(A)P(B_2)$$

Let $B_3 =$ "the sum of the two dice is 9." This time the occurrence of A enhances the probability of B_3, i.e., $P(B_3|A) = 1/6 > 4/36 = P(B_3)$, so the two events are not independent. To check that this claim using (A.6), we note that (A.4) implies

$$P(A \cap B_3) = P(A)P(B_3|A) > P(A)P(B_3)$$

Let $B_4 =$ "the sum of the two dice is 7." Somewhat surprisingly, A and B_4 are independent. To check this from (A.6), we note that $P(B_4) = 6/36$ and $A \cap B_4 = \{(4,3)\}$ has probability 1/36, so

$$P(A \cap B_3) = \frac{1}{36} = \frac{1}{6} \cdot \frac{6}{36} = P(A)P(B_3)$$

There are two ways of extending the definition of independence to more than two events.

$A_1, \ldots, A_n$ are said to be **pairwise independent** if for each $i \neq j$, $P(A_i \cap A_j) = P(A_i)P(A_j)$, that is, each pair is independent.

$A_1, \ldots, A_n$ are said to be **independent** if for any $1 \le i_1 < i_2 < \ldots < i_k \le n$ we have

$$P(A_{i_1} \cap \ldots \cap A_{i_k}) = P(A_{i_1}) \cdots P(A_{i_k})$$

If we flip n coins and let $A_i = $ "the ith coin shows Heads," then the A_i are independent since $P(A_i) = 1/2$ and for any choice of indices $1 \le i_1 < i_2 < \ldots < i_k \le n$ we have $P(A_{i_1} \cap \ldots \cap A_{i_k}) = 1/2^k$. Our next example shows that events can be pairwise independent but not independent.

Example A.7. Flip three coins. Let $A = $ "the first and second coins are the same," $B = $ "the second and third coins are the same," and $C = $ "the third and first coins are the same." Clearly $P(A) = P(B) = P(C) = 1/2$. The intersection of any two of these events is

$$A \cap B = B \cap C = C \cap A = \{HHH, TTT\}$$

an event of probability 1/4. From this it follows that

$$P(A \cap B) = \frac{1}{4} = \frac{1}{2} \cdot \frac{1}{2} = P(A)P(B)$$

i.e., A and B are independent. Similarly, B and C are independent and C and A are independent; so A, B, and C are pairwise independent. The three events A, B, and C are not independent, however, since $A \cap B \cap C = \{HHH, TTT\}$ and hence

$$P(A \cap B \cap C) = \frac{1}{4} \ne \left(\frac{1}{2}\right)^3 = P(A)P(B)P(C)$$

The last example is somewhat unusual. However, the moral of the story is that to show several events are independent, you have to check more than just that each pair is independent.

A.2   Random Variables, Distributions

Formally, a **random variable** is a real-valued function defined on the sample space. However, in most cases the sample space is usually not visible, so we describe the random variables by giving their distributions. In the **discrete case** where the random variable can take on a finite or countably infinite set of values this is usually done using the **probability function**. That is, we give $P(X = x)$ for each value of x for which $P(X = x) > 0$.

Example A.8 (Binomial Distribution). If we perform an experiment n times and on each trial there is a probability p of success, then the number of successes S_n has

$$P(S_n = k) = \binom{n}{k} p^k (1 - p)^{n-k} \qquad \text{for } k = 0, \dots, n$$

In words, S_n has a binomial distribution with parameters n and p, a phrase we will abbreviate as $S_n = \text{binomial}(n, p)$.

Example A.9 (Geometric Distribution). If we repeat an experiment with probability p of success until a success occurs, then the number of trials required, N, has

$$P(N = n) = (1 - p)^{n-1} p \qquad \text{for } n = 1, 2, \dots$$

In words, N has a geometric distribution with parameter p, a phrase we will abbreviate as $N = \text{geometric}(p)$.

Example A.10 (Poisson Distribution). X is said to have a Poisson distribution with parameter $\lambda > 0$, or $X = \text{Poisson}(\lambda)$ if

$$P(X = k) = e^{-\lambda} \frac{\lambda^k}{k!} \quad \text{for } k = 0, 1, 2, \dots$$

To see that this is a probability function we recall

$$e^x = \sum_{k=0}^{\infty} \frac{x^k}{k!} \tag{A.7}$$

so the proposed probabilities are nonnegative and sum to 1.

In many situations random variables can take any value on the real line or in a certain subset of the real line. For concrete examples, consider the height or weight of a person chosen at random or the time it takes a person to drive from Los Angeles to San Francisco. A random variable X is said to have a **continuous distribution** with **density function** f if for all $a \leq b$ we have

$$P(a \leq X \leq b) = \int_a^b f(x)\, dx \tag{A.8}$$

Geometrically, $P(a \leq X \leq b)$ is the area under the curve f between a and b.

In order for $P(a \leq X \leq b)$ to be nonnegative for all a and b and for $P(-\infty < X < \infty) = 1$ we must have

$$f(x) \geq 0 \quad \text{and} \quad \int_{-\infty}^{\infty} f(x)\, dx = 1 \tag{A.9}$$

Any function f that satisfies (A.9) is said to be a **density function**. We will now define three of the most important density functions.

Example A.11 (Uniform Distribution on (a,b)).

$$f(x) = \begin{cases} 1/(b-a) & a < x < b \\ 0 & \text{otherwise} \end{cases}$$

The idea here is that we are picking a value "at random" from (a, b). That is, values outside the interval are impossible, and all those inside have the same probability density. Note that the last property implies $f(x) = c$ for $a < x < b$. In this case the integral is $c(b-a)$, so we must pick $c = 1/(b-a)$.

Example A.12 (Exponential Distribution).

$$f(x) = \begin{cases} \lambda e^{-\lambda x} & x \geq 0 \\ 0 & \text{otherwise} \end{cases}$$

Here $\lambda > 0$ is a parameter. To check that this is a density function, we note that

$$\int_0^\infty \lambda e^{-\lambda x}\, dx = -e^{-\lambda x}\big|_0^\infty = 0 - (-1) = 1$$

In a first course in probability, the next example is the star of the show. However, it will have only a minor role here.

Example A.13 (Normal Distribution).

$$f(x) = (2\pi)^{-1/2} e^{-x^2/2}$$

Since there is no closed form expression for the antiderivative of f, it takes some ingenuity to check that this is a probability density. Those details are not important here, so we will ignore them.

Any random variable (discrete, continuous, or in between) has a **distribution function** defined by $F(x) = P(X \leq x)$. If X has a density function $f(x)$, then

$$F(x) = P(-\infty < X \leq x) = \int_{-\infty}^x f(y)\, dy$$

That is, F is an antiderivative of f.

One of the reasons for computing the distribution function is explained by the next formula. If $a < b$, then $\{X \leq b\} = \{X \leq a\} \cup \{a < X \leq b\}$ with the two sets on the right-hand side disjoint so

$$P(X \leq b) = P(X \leq a) + P(a < X \leq b)$$

or, rearranging,

$$P(a < X \leq b) = P(X \leq b) - P(X \leq a) = F(b) - F(a) \qquad \text{(A.10)}$$

The last formula is valid for any random variable. When X has density function f, it says that

$$\int_a^b f(x)\,dx = F(b) - F(a)$$

i.e., the integral can be evaluated by taking the difference of the antiderivative at the two endpoints.

To see what distribution functions look like, and to explain the use of (A.10), we return to our examples.

Example A.14 (Uniform Distribution). $f(x) = 1/(b-a)$ for $a < x < b$.

$$F(x) = \begin{cases} 0 & x \le a \\ (x-a)/(b-a) & a \le x \le b \\ 1 & x \ge b \end{cases}$$

To check this, note that $P(a < X < b) = 1$ so $P(X \le x) = 1$ when $x \ge b$ and $P(X \le x) = 0$ when $x \le a$. For $a \le x \le b$ we compute

$$P(X \le x) = \int_{-\infty}^x f(y)\,dy = \int_a^x \frac{1}{b-a}\,dy = \frac{x-a}{b-a}$$

In the most important special case $a = 0$, $b = 1$ we have $F(x) = x$ for $0 \le x \le 1$.

Example A.15 (Exponential Distribution). $f(x) = \lambda e^{-\lambda x}$ for $x \ge 0$.

$$F(x) = \begin{cases} 0 & x \le 0 \\ 1 - e^{-\lambda x} & x \ge 0 \end{cases}$$

The first line of the answer is easy to see. Since $P(X > 0) = 1$, we have $P(X \le x) = 0$ for $x \le 0$. For $x \ge 0$ we compute

$$P(X \le x) = \int_0^x \lambda e^{-\lambda y}\,dy = -e^{-\lambda y}\big|_0^x = 1 - e^{-\lambda x}$$

In many situations we need to know the relationship between several random variables $X_1, \ldots, X_n$. If the X_i are discrete random variables then this is easy, we simply give the probability function that specifies the value of

$$P(X_1 = x_1, \ldots, X_n = x_n)$$

whenever this is positive. When the individual random variables have continuous distributions this is described by giving the **joint density function** which has the interpretation that

$$P((X_1, \ldots, X_n) \in A) = \int \cdots \int_A f(x_1, \ldots, x_n)\,dx_1 \ldots dx_n$$

By analogy with (A.9) we must require that $f(x_1, \ldots, x_n) \geq 0$ and

$$\int \cdots \int f(x_1, \ldots, x_n)\, dx_1 \ldots dx_n = 1$$

Having introduced the joint distribution of n random variables, we will for simplicity restrict our attention for the rest of the section to $n = 2$. The first question we will confront is: "Given the joint distribution of (X, Y), how do we recover the distributions of X and Y?" In the discrete case this is easy. The **marginal distributions** of X and Y are given by

$$P(X = x) = \sum_y P(X = x, Y = y)$$

$$P(Y = y) = \sum_x P(X = x, Y = y) \qquad\qquad (A.11)$$

To explain the first formula in words, if $X = x$, then Y will take on some value y, so to find $P(X = x)$ we sum the probabilities of the disjoint events $\{X = x, Y = y\}$ over all the values of y.

Formula (A.11) generalizes in a straightforward way to continuous distributions: we replace the sum by an integral and the probability functions by density functions. If X and Y have joint density $f_{X,Y}(x, y)$, then the **marginal densities** of X and Y are given by

$$f_X(x) = \int f_{X,Y}(x, y)\, dy$$

$$f_Y(y) = \int f_{X,Y}(x, y)\, dx \qquad\qquad (A.12)$$

The verbal explanation of the first formula is similar to that of the discrete case: if $X = x$, then Y will take on some value y, so to find $f_X(x)$ we integrate the joint density $f_{X,Y}(x, y)$ over all possible values of y.

Two random variables are said to be **independent** if for any two sets A and B we have

$$P(X \in A, Y \in B) = P(X \in A)P(Y \in B) \qquad\qquad (A.13)$$

In the discrete case, (A.13) is equivalent to

$$P(X = x, Y = y) = P(X = x)P(Y = y) \qquad\qquad (A.14)$$

for all x and y. The condition for independence is exactly the same in the continuous case: the joint distribution is the product of the marginal densities.

$$f_{X,Y}(x, y) = f_X(x)f_Y(y) \qquad\qquad (A.15)$$

The notions of independence extend in a straightforward way to n random variables: the joint probability or probability density is the product of the marginals.

Two important consequences of independence are

Theorem A.1. *If $X_1, \ldots X_n$ are independent, then*

$$E(X_1 \cdots X_n) = EX_1 \cdots EX_n$$

Theorem A.2. *If $X_1, \ldots X_n$ are independent and $n_1 < \ldots < n_k \le n$, then*

$$h_1(X_1, \ldots X_{n_1}), h_2(X_{n_1+1}, \ldots X_{n_2}), \ldots h_k(X_{n_{k-1}+1}, \ldots X_{n_k})$$

are independent.

In words, the second result says that functions of disjoint sets of independent random variables are independent.

Our last topic in this section is the distribution of $X + Y$ when X and Y are independent. In the discrete case this is easy:

$$P(X + Y = z) = \sum_x P(X = x)P(Y = z - x) \tag{A.16}$$

To see the first equality, note that if the sum is z then X must take on some value x and Y must be $z - x$. The first equality is valid for any random variables. The second holds since we have supposed X and Y are independent.

Example A.16. If $X = $ binomial(n, p) and $Y = $ binomial(m, p) are independent, then $X + Y = $ binomial$(n + m, p)$.

Proof by Direct Computation.

$$P(X + Y = i) = \sum_{j=0}^{i} \binom{n}{j} p^j (1-p)^{n-j} \cdot \binom{m}{i-j} p^{i-j} (1-p)^{m-i+j}$$

$$= p^i (1-p)^{n+m-i} \sum_{j=0}^{i} \binom{n}{j} \cdot \binom{m}{i-j}$$

$$= \binom{n+m}{i} p^i (1-p)^{n+m-i}$$

The last equality follows from the fact that if we pick i individuals from a group of n boys and m girls, which can be done in $\binom{n+m}{i}$ ways, then we must have j boys and $i - j$ girls for some j with $0 \le j \le i$. $\qquad\square$

Much Easier Proof Consider a sequence of $n + m$ independent trials. Let X be the number of successes in the first n trials and Y be the number of successes in the last m. By (2.13), X and Y independent. Clearly their sum is binomial(n, p). $\qquad\square$

Formula (A.16) generalizes in the usual way to continuous distributions: regard the probabilities as density functions and replace the sum by an integral.

$$f_{X+Y}(z) = \int f_X(x) f_Y(z-x)\, dx \qquad (A.17)$$

Example A.17. Let U and V be independent and uniform on $(0,1)$. Compute the density function for $U + V$.

Solution. If $U + V = x$ with $0 \le x \le 1$, then we must have $U \le x$ so that $V \ge 0$. Recalling that we must also have $U \ge 0$

$$f_{U+V}(x) = \int_0^x 1 \cdot 1\, du = x \quad \text{when } 0 \le x \le 1$$

If $U+V = x$ with $1 \le x \le 2$, then we must have $U \ge x-1$ so that $V \le 1$. Recalling that we must also have $U \le 1$,

$$f_{U+V}(x) = \int_{x-1}^1 1 \cdot 1\, du = 2 - x \quad \text{when } 1 \le x \le 2$$

Combining the two formulas we see that the density function for the sum is triangular. It starts at 0 at 0, increases linearly with rate 1 until it reaches the value of 1 at $x = 1$, then it decreases linearly back to 0 at $x = 2$. $\square$

A.3   Expected Value, Moments

If X has a discrete distribution, then the **expected value** of $h(X)$ is

$$Eh(X) = \sum_x h(x) P(X = x) \qquad (A.18)$$

When $h(x) = x$ this reduces to EX, the expected value, or **mean of** X, a quantity that is often denoted by μ or sometimes μ_X to emphasize the random variable being considered. When $h(x) = x^k$, $Eh(X) = EX^k$ is the *k*th **moment**. When $h(x) = (x - EX)^2$,

$$Eh(X) = E(X - EX)^2 = EX^2 - (EX)^2$$

is called the **variance** of X. It is often denoted by $\operatorname{var}(X)$ or σ_X^2. The variance is a measure of how spread out the distribution is. However, if X has the units of feet, then the variance has units of feet^2, so the **standard deviation** $\sigma(X) = \sqrt{\operatorname{var}(X)}$, which has again the units of feet, gives a better idea of the "typical" deviation from the mean than the variance does.

Example A.18 (Roll One Die). $P(X = x) = 1/6$ for $x = 1, 2, 3, 4, 5, 6$ so

$$EX = (1 + 2 + 3 + 4 + 5 + 6) \cdot \frac{1}{6} = \frac{21}{6} = 3.5$$

In this case the expected value is just the average of the six possible values.

$$EX^2 = (1^2 + 2^2 + 3^2 + 4^2 + 5^2 + 6^2) \cdot \frac{1}{6} = \frac{91}{6}$$

so the variance is $91/6 - 49/4 = 70/24$. Taking the square root we see that the standard deviation is 1.71. The three possible deviations, in the sense of $|X - EX|$, are 0.5, 1.5, and 2.5 with probability 1/3 each, so 1.71 is indeed a reasonable approximation for the typical deviation from the mean.

Example A.19 (Geometric Distribution).
Starting with the sum of the geometric series

$$(1 - \theta)^{-1} = \sum_{n=0}^{\infty} \theta^n$$

and then differentiating twice and discarding terms that are 0 gives

$$(1 - \theta)^{-2} = \sum_{n=1}^{\infty} n\theta^{n-1} \quad \text{and} \quad 2(1 - \theta)^{-3} = \sum_{n=2}^{\infty} n(n-1)\theta^{n-2}$$

Using these with $\theta = 1 - p$, we see that

$$EN = \sum_{n=1}^{\infty} n(1-p)^{n-1}p = p/p^2 = \frac{1}{p}$$

$$EN(N-1) = \sum_{n=2}^{\infty} n(n-1)(1-p)^{n-1}p = 2p^{-3}(1-p)p = \frac{2(1-p)}{p^2}$$

and hence

$$\text{var}\,(N) = EN(N-1) + EN - (EN)^2$$

$$= \frac{2(1-p)}{p^2} + \frac{p}{p^2} - \frac{1}{p^2} = \frac{(1-p)}{p^2}$$

The definition of expected value generalizes in the usual way to continuous random variables. We replace the probability function by the density function and the sum by an integral

$$Eh(X) = \int h(x)f_X(x)\,dx \tag{A.19}$$

Example A.20 (Uniform Distribution on (a,b)). Suppose X has density function $f_X(x) = 1/(b-a)$ for $a < x < b$ and 0 otherwise. In this case

$$EX = \int_a^b \frac{x}{b-a}\,dx = \frac{b^2 - a^2}{2(b-a)} = \frac{(b+a)}{2}$$

since $b^2 - a^2 = (b-a)(b+a)$. Notice that $(b+a)/2$ is the midpoint of the interval and hence the natural choice for the average value of X. A little more calculus gives

$$EX^2 = \int_a^b \frac{x^2}{b-a}\,dx = \frac{b^3 - a^3}{3(b-a)} = \frac{b^2 + ba + a^2}{3}$$

since $b^3 - a^3 = (b-a)(b^2 + ba + a^2)$. Squaring our formula for EX gives $(EX)^2 = (b^2 + 2ab + a^2)/4$, so

$$\text{var}(X) = (b^2 - 2ab + a^2)/12 = (b-a)^2/12$$

To help explain the answers we have found in the last two example we use

Theorem A.3. *If c is a real number, then*

(a) $E(X + c) = EX + c$ (b) $\text{var}(X + c) = \text{var}(X)$
(c) $E(cX) = cEX$ (d) $\text{var}(cX) = c^2\,\text{var}(X)$

Uniform Distribution on (a,b) If X is uniform on $[(a-b)/2, (b-a)/2]$, then $EX = 0$ by symmetry. If $c = (a+b)/2$, then $Y = X + c$ is uniform on $[a, b]$, so it follows from (a) and (b) of Theorem A.3 that

$$EY = EX + c = (a+b)/2 \qquad \text{var}(Y) = \text{var}(X)$$

From the second formula we see that the variance of the uniform distribution will only depend on the length of the interval. To see that it will be a multiple of $(b-a)^2$ note that $Z = X/(b-a)$ is uniform on $[-1/2, 1/2]$ and then use part (d) of Theorem A.3 to conclude $\text{var}(X) = (b-a)^2\,\text{var}(Z)$. Of course one needs calculus to conclude that $\text{var}(Z) = 1/12$.

Generating functions will be used at several points in the text. If $p_k = P(X = k)$ is the distribution of X, then the generating function is $\phi(x) = \sum_{k=0}^{\infty} p_k x^k$. $\phi(1) = \sum_{k=0}^{\infty} p_k = 1$. Differentiating (and not worrying about the detail of interchanging the sum and the integral) we have

$$\phi'(x) = \sum_{k=1}^{\infty} k p_k x^{k-1} \qquad \phi'(1) = EX$$

or in general after m derivatives

$$\phi^{(m)}(x) = \sum_{k=m}^{\infty} k(k-1)\cdots(k-m+1)p_k x^{k-1}$$

$$\phi^{(m)}(1) = E[X(X-1)\cdots(X-m+1)]$$

Example A.21 (Poisson Distribution). $P(X = k) = e^{-\lambda k}\lambda^k/k!$. The generating function is

$$\phi(x) = \sum_{k=0}^{\infty} e^{-\lambda k}\frac{\lambda^k x^k}{k!} = \exp(-\lambda + \lambda x)$$

Differentiating m times we have

$$\phi^{(m)}(x) = \lambda^m \exp(-\lambda(1-x))$$

and $E[X(X-1)\cdots(X-m+1)] = \lambda^m$. From this we see that $EX = \lambda$ and reasoning as we did for the geometric

$$\text{var}(X) = EX(X-1) + EX - (EN)^2 = \lambda^2 + \lambda - \lambda^2 = \lambda$$

The next two results give important properties of expected value and variance.

Theorem A.4. *If $X_1,\ldots,X_n$ are any random variables, then*

$$E(X_1 + \cdots + X_n) = EX_1 + \cdots + EX_n$$

Theorem A.5. *If $X_1,\ldots,X_n$ are independent, then*

$$\text{var}(X_1 + \cdots + X_n) = \text{var}(X_1) + \cdots + \text{var}(X_n)$$

Theorem A.6. *If $X_1,\ldots,X_n$ are independent and have a distribution with generating function $\phi(x)$, then the generating function of the sum is*

$$E(x^{S_n}) = \phi(x)^n$$

To illustrate the use of these properties we consider the

Example A.22 (Binomial Distribution). If we perform an experiment n times and on each trial there is a probability p of success, then the number of successes S_n has

$$P(S_n = k) = \binom{n}{k}p^k(1-p)^{n-k} \qquad \text{for } k = 0,\ldots,n$$

To compute the mean and variance we begin with the case $n = 1$, which is called the Bernoulli distribution Writing X instead of S_1 to simplify notation, we have $P(X = 1) = p$ and $P(X = 0) = 1 - p$, so

$$EX = p \cdot 1 + (1 - p) \cdot 0 = p$$
$$EX^2 = p \cdot 1^2 + (1 - p) \cdot 0^2 = p$$
$$\text{var}(X) = EX^2 - (EX)^2 = p - p^2 = p(1 - p)$$

To compute the mean and variance of S_n, we observe that if $X_1, \ldots, X_n$ are independent and have the same distribution as X, then $X_1 + \cdots + X_n$ has the same distribution as S_n. Intuitively, this holds since $X_i = 1$ means one success on the ith trial so the sum counts the total number of success. Using Theorems A.4 and A.5, we have

$$ES_n = nEX = np \qquad \text{var}(S_n) = n \, \text{var}(X) = np(1 - p)$$

As for the generating function. When $n = 1$ it is $(1 - p + px)$ so by Theorem A.6 it is

$$(1 - p + px)^n$$

in general. If we set $p = \lambda/n$ and let $n \to \infty$, then

$$\left(1 - \frac{\lambda}{n}(1 - x)\right)^n \to \exp(-\lambda(1 - x))$$

the generating function of the Poisson.

In some cases an alternate approach to computing the expected value of X is useful. In the discrete case the formula is

Theorem A.7. *If $X \geq 0$ is integer valued, then*

$$EX = \sum_{k=1}^{\infty} P(X \geq k) \qquad\qquad (A.20)$$

Proof. Let $1_{\{X \geq k\}}$ denote the random variable that is 1 if $X \geq k$ and 0 otherwise. It is easy to see that

$$X = \sum_{k=1}^{\infty} 1_{\{X \geq k\}}.$$

Taking expected values and noticing $E1_{\{X \geq k\}} = P(X \geq k)$ gives

$$EX = \sum_{k=1}^{\infty} P(X \geq k)$$

which proves the desired result.   □

The analogous result which holds in general is:

Theorem A.8. *Let $X \geq 0$. Let H be a differentiable nondecreasing function with $H(0) = 0$. Then*

$$EH(X) = \int_0^{\infty} H'(t)P(X > t)\, dt$$

Proof. We assume H is nondecreasing only to make sure that the integral exists. (It may be ∞.) Introducing the indicator $1_{\{X>t\}}$ that is 1 if $X > t$ and 0 otherwise, we have

$$\int_0^{\infty} H'(t)1_{\{X>t\}} = \int_0^X H'(t)\, dt = H(X)$$

and taking expected value gives the desired result.   □

Taking $H(x) = x^p$ with $p > 0$ we have

$$EX^p = \int_0^{\infty} pt^{p-1}P(X > t)\, dt \qquad (A.21)$$

When $p = 1$ this becomes

$$EX = \int_0^{\infty} P(X > t)\, dt \qquad (A.22)$$

the analogue to (A.21) in the discrete case is

$$EX^p = \sum_{k=1}^{\infty} (k^p - (k-1)^p)P(X \geq k) \qquad (A.23)$$

When $p = 2$ this becomes

$$EX^2 = \sum_{k=1}^{\infty} (2k - 1)P(X \geq k) \qquad (A.24)$$

To state our final useful fact, recall that ϕ is convex if for all x, y and $\lambda \in (0, 1)$

$$\phi(\lambda x + (1 - \lambda)y) \leq \lambda\phi(x) + (1 - \lambda)\phi(y)$$

For a smooth function this is equivalent to ϕ' is nondecreasing or $\phi'' \geq 0$.

Theorem A.9. *If ϕ is convex, then $E\phi(X) \geq \phi(EX)$.*

Proof. There is a linear function $\ell(y) = \phi(EX) + c(y - EX)$ so that $\ell(y) \leq \phi(y)$ for all y. If one accepts this fact the proof is easy. Replacing y by X and taking expected value we have

$$E\phi(X) \geq E\ell(X) = \phi(EX)$$

since $E(X - EX) = 0$. To prove the fact we note that for any z, as $h \downarrow 0$

$$\frac{\phi(z+h) - \phi(z)}{h} \downarrow c_+ \qquad \frac{\phi(z) - \phi(z-h)}{h} \uparrow c_-$$

Taking $z = EX$ and $c \in [c_-, c_+]$ gives the desired linear function.   □

A.4   Integration to the Limit

This is not review material since it is material usually covered in a graduate probability class. Nonetheless it is important for giving a rigorous treatment of some topics in chapter so we collect results here for reference. The main question to be considered here is: when does the convergence of random variable X_n to a limit X imply that $EX_n \to EX$. Our purpose here is simply to state clearly three sufficient conditions. We begin with a counterexample.

Example A.23. Consider $\{1, 2, 3, \ldots\}$ with probability measure $P(\{k\}) = 2^{-k}$. Suppose we have random variables $X_n(n) = 2^n$ and $X_n = 0$ otherwise. Then $X_n \to 0$ but each X_n has $EX_n = 1$.

The next three results give sufficient conditions for $EX_n \to EX$. The proofs can be found in any book on measure-theoretic probability, so we do not go into details.

Theorem A.10. *If there is a constant K so that $|X_n| \leq K$ for all n, then $EX_n \to EX$.*

Theorem A.11. *If $X_n \geq 0$ and X_n is increasing in n, then $EX_n \uparrow EX$.*

Theorem A.12. *If $|X_n| \leq Y$ and $EY < \infty$, then $EX_n \to EX$.*

References

Allen, Linda J.S. (2003) *An Introduction to Stochastic Processes with Applications to Biology.* Pearson Prentice Hall, Upper Saddle River, NJ

Athreya, K.B. and Ney, P.E. (1972) *Branching Processes.* Springer-Verlag, New York

Bailey, N.T.J. (1964) *The Elements of Stochastic Processes: With Applications to the Natural Sciences.* John Wiley and Sons.

Barbour, A.D., Holst, L., and Janson, S. (1992) *Poisson Approximation.* Oxford U. Press

Bhattacharya, R.N. and Waymire, C. (1990) *Stochastic Processes with Applications.* John Wiley and Sons, New York

Chung, K.L. (1967) *Markov Chains with Stationary Transition Probabilities.* Second edition, Springer-Verlag, New York

Cox, D.R. and Miller, H.D. (1965) *The Theory of Stochastic Processes.* Methuen & Co, Ltd., London

Feller, W. (1968) *An Introduction to Probability Theory and its Applications.* Third edition, John Wiley and Sons, New York

Geman, S. and Geman, D. (1984) Stochastic relaxation, Gibbs distributions and the Bayesian restoration ofimages. *IEEE Transactions on Pattern Analysis and Machine Intelligence.* 6, 721–741

Gliovich, T., Vallone, R., and Tversky, A. (1985) The Hot Hand in Basketball: On the Misperception of Random Sequences. *Cognitive Psychology.* 17, 295–314

Goodman, V., and Stampfli, J. (2009) *The Mathematics of Finance: Modeling and Hedging.* American Math Society. Providence, RI

Hammersley, J.M., and D.C. Handscomb, D.C. (1984) *Monte Carlo Methods.* Chapman and Hall, London

Hasegawa, M., Kishino, H., Yano, T. (1985) *Dating of human-ape splitting by a molecular clock of mitochondrial DNA.* Journal of Molecular Evolution. 22(2):160–174

Hastings, W.K. (1970) Monte Carlo sampling methods using Markov chains and their applications. *Biometrika* 57, 97–109

Hoel, P.G., Port, S.C. and Stone, C.J. (1972) *Introduction to Stochastic Processes.* Houghton-Mifflin, Boston, MA

Hull, J.C. (1997) *Options, Futures, and Other Derivatives.* Prentice Hall, Upper Saddle River, NJ

Jagers, P. (1975) *Branching Processes with Biological Applications.* John Wiley and Sons, New York

Karlin, S. and Taylor, H.M. (1975) *A First Course in Stochastic Processes.* Academic Press, New York

Kelly, F. (1979) *Reversibility and Stochastic Networks.* John Wiley and Sons, New York

© Springer International Publishing Switzerland 2016

R. Durrett, *Essentials of Stochastic Processes*, Springer Texts in Statistics,
DOI 10.1007/978-3-319-45614-0

Kesten, H., and Stigum, B. (1966) A limit theorem for multi-dimensional Galton-Watson processes. *Ann. Math. Statist.* 37, 1211–1223

Kingman, J.F.C. (1993) *Poisson Processes.* Oxford U. Press

Kirkpatrick, S., Gelatt, Jr., C.D., and Vecchi, M.PP. (1983) Optimizing by simulated annealing. *Science.* 220, 671–680

Lange, K. (2003) *Applied Probability.* Springer-Verlag, New York

Lawler, G.F. (2006) *Introduction to Stochastic Processes.* Second edition. Chapman and Hall/CRC, Boca Raton, FL

Metropolis, N., Rosenbluth, A., Rosenbluth, M., Teller, A. and Teller, E. (1953) Equation of state computation by fast computing machines. *J. Chemical Physics.* 21, 1087–1092

Neveu, J. (1975) *Discrete Parameter Martingales.* North Holland, Amsterdam

Resnick, S.I. (1992) *Adventures in Stochastic Processes.* Birkhäuser, Boston

Ross, S.M. (2007) Introduction to Probability Models. Ninth edition. Academic Press, New York

Shreve, S.E. (2004) *Stochastic Calculus for Finance, I.* Springer, New York

Takacs, L. (1962) *Introduction to the Theory of Queues.* Oxford U Press

Tierney, L. (1994) Markov chains for exploring posterior distributions. *Annals of Statistics.* 22,, 1701–1762

Whittle, P. (1992) *Probability via Expectation.* Third Edition. Springer, New York

Index

A

American option
    definition, 237
    vs. European option, 239–240
    optimal strategy, 238–239
    risk neutral probabilities, 238, 240
Arbitrage opportunity, 224–225

B

Bayes' formula, 254
Binomial model
    callback options, 233
    knockout options, 236–237
    N period model, 229–232
    one period case, 227–229
    put-call parity, 235–236
    put option, 234–235
    risk neutral probabilities, 232
Black–Scholes formula
    central limit theorem, 241–242, 244
    exponential growth, 242
    mean and variance, 244
    moment generating function, 243
    partial differential equation, 244–245
    risk neutral measure, 244
    stock price model, 242
    volatility, 242
Busy periods, 177, 179

C

Callback options, 233
Chapman–Kolmogorov equation, 148
Classification of states

expected values, 19, 20
finite state Markov chain, 17
Gambler's Ruin, 15
irreducible closed sets, 18
nonnegative integer valued random
        variable, 19
seven-state chain, 17–18
social mobility, 16
stopping time (T), 13
strong Markov property, 13–14, 18
transient and recurrent states, 16, 20
Compound Poisson processes, 106–108
Conditional expectation
    disjoint union, 203
    expected value, 202
    indicator function, 201
    Jensen's inequality, 202
Conditional probability, 9, 11, 21, 253–255
Conditioning, 113–114, 120–122
Continuous time Markov chains
    branching process, 150
    Chapman–Kolmogorov equation, 148
    computer simulation, 151–152
    definitions, 147
    exit distributions
        branching process, 170–172
        embedded jump chain, 170
        office hours, 172–173
    exit times, 173–176
    formal construction, 150
    jump rate, 148–149
    limiting behavior
        detailed balance condition, 167–170
        Duke basketball, 165–167
        irreducible, 163–164

© Springer International Publishing Switzerland 2016
R. Durrett, *Essentials of Stochastic Processes*, Springer Texts in Statistics,
DOI 10.1007/978-3-319-45614-0

Printed in the United States
By Bookmasters